THE CERTIFICATION SERIES

US SAILING®

Coastal Navigation

The national standard for quality sailing instruction

 ISBN 1-882502-34-5. Printed in the United States of America
UNITED STATES SAILING ASSOCIATION P.O. Box 1260, 15 Maritime Drive, Portsmouth, RI 02871-6015

Acknowledgments

This fourth book of US SAILING's Certification Series once again brings together our special design team of Mark Smith, Kim Downing and Rob Eckhardt with the experienced navigation writer, Tom Cunliffe. We would like to give special recognition to David Forbes and Phil Shull who developed the passage plan in the last chapter and contributed technical input and examples, and to Joanne M. Dorval who coordinated and expedited the printing of The Certification Series. In addition, we would like to thank the students who used the test and evaluation edition and provided invaluable feedback as well as the sailing schools and volunteers who were resources for guidance, review, and advice. These include Tim Broderick, Steve Colgate, Richard Johnson, George Moffett, James Muldoon, Timmy Larr, Pete Niewieroski, Tyler Pierce, Anthony Sandberg, Barry Swackhamer, Susie Trotman and Ray Wichmann. Last, but not least, the foresight and financial support of Sail America and US SAILING's Officers and Board have been instrumental to the production of the entire certification series.

Cover photo courtesy of *SAIL* Magazine

Tom Cunliffe, *Author*
Tom brings his vas experience as professional sailor and RYA Yachtmaster Examiner to his writing. His articles appear in major periodicals and his books include a four book navigation series, *Easy on the Helm, Cunliffe on Cruising, Topsail and Battleaxe* and *Hand Reef and Steer*. These last two won BEST BOOK OF THE SEA prizes. He also finds time to cruise in his 35-ton Edwardian gaff cutter with his wife Ros and daughter Hannah to such diverse places as Brazil, Greenland, the Caribbean, the U.S., Labrador and Russia.

Mark Smith, *Designer*
A lifelong sailor, graphic designer, editor and illustrator, Mark is currently Creative Director for North Sails. Mark was editorial and art director for *Yacht Racing / Cruising* magazine (now *Sailing World*) from 1970-83, editor and publisher of *Sailor* magazine from 1984-86, and editor and art director of *American Sailor* from 1987-89. His works include design and illustration for the *Annapolis Book of Seamanship* by John Rousmaniere and published by Simon and Schuster. Mark lives in Rowayton, CT with his wife Tina and daughters Stephanie, Natalie and Cristina.

Kim Downing, *Illustrator*
Kim grew up in the Midwest doing two things, sailing and drawing, so it's only natural that his two favorite pastimes should come together in the production of this book. Kim is the proprietor of MAGAZINE ART and provides technical illustrations to magazine and book publishers. He recently helped his father complete the building of a custom 30-foot sailboat and enjoys racing and daysailing his own boat with his wife and two children.

Rob Eckhardt, *Illustrator*
A graphic design professional, Rob is currently on the staff of *SAIL* Magazine and has many years of experience as a designer for advertising agencies, publications and his own business clients. He is a graduate of the Rochester Institute of Technology, Rochester, NY. Rob began sailing dinghies as a youngster and currently enjoys one-design racing and coastal cruising.

Foreword

Sailing and navigation go hand-in-hand. Having spent much of my adult life sailing, I have also spent a great deal of time learning and perfecting my navigating skills. Much of what I've achieved in this wonderful sport has been the direct result of navigation courses I have taken (as well as many lessons learned through experience!). It has all helped me to be considered one of the world's top female navigators.

Navigation is a never-ending challenge, but the rewards and satisfaction gained from meeting that challenge are difficult to top. Whether you cruise the Gulf of Mexico, the Pacific Islands or the French Riviera, sound navigation is the foundation of any safe, enjoyable cruising experience. In racing as well, it is precise, proactive navigation that often defines the difference between winning and losing.

In the past, becoming a navigator meant reading books and taking courses that were often difficult, time-consuming, and frustrating for many a novice. With a book as clear, understandable and enjoyable as this, all that has changed. The brotherhood (and sisterhood) of navigators is about to get a whole lot bigger. Welcome!

Christy Crawford

Christy Steinman Crawford

Christy Crawford is one of the sailing world's most accomplished navigators. She has navigated in prominent offshore racing events around the world, including the Southern Ocean Racing Conference (8 times), Kenwood Cup (4 times), Admiral's Cup (3 times) and Fastnet Race (3 times). Christy was the first woman to serve as navigator on an America's Cup team, sailing aboard Dennis Conner's Freedom *campaign in 1983. Since 1988 she has cruised worldwide with her husband and two children.*

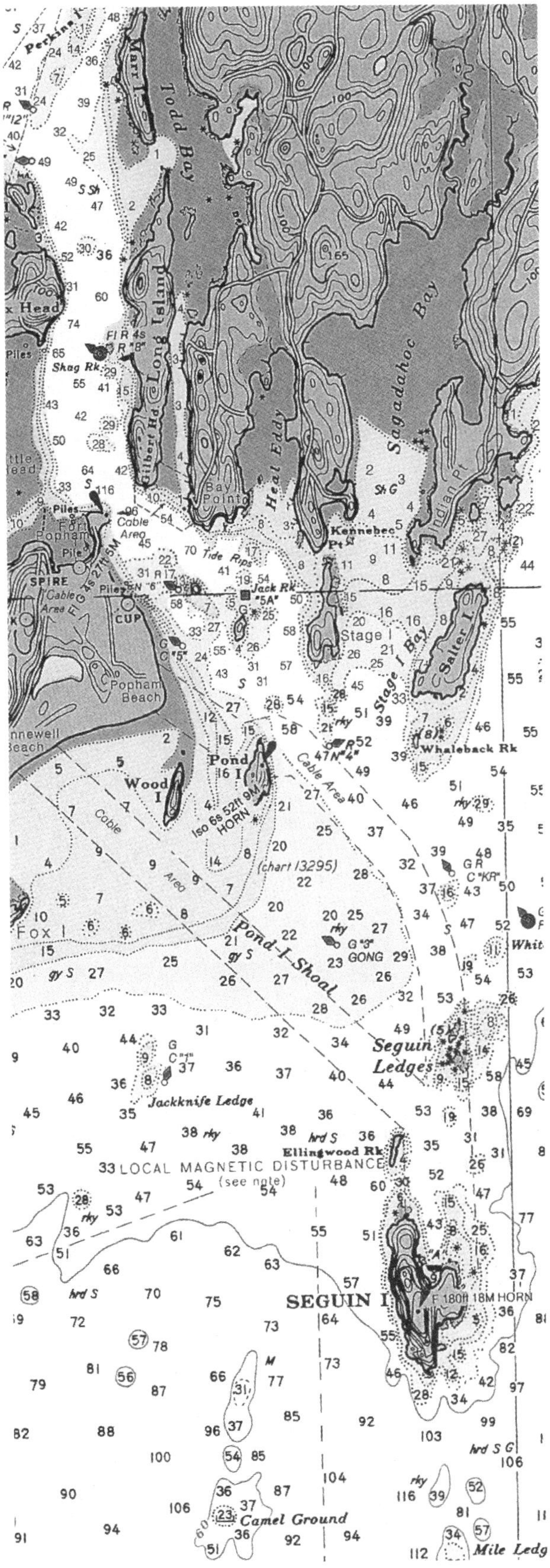
Todd Bay
Long Island
Sagadahoc Bay
Heal Eddy
Gilbert Hd
Bay Point
Kennebec Pt
Indian Pt
Shag Rk
Popham Beach
Jack Rk
Stage I
Stage I Bay
Salter I
Whaleback Rk
Wood I
Pond I
Pond-I-Shoal
Fox I
Seguin Ledges
Jackknife Ledge
Ellingwood Rk
LOCAL MAGNETIC DISTURBANCE
(see note)
SEGUIN I
Camel Ground
Mile Ledg
Cable Area
Tide Rips
(chart 13295)
SPIRE
CUP

CHAPTER 1 Introduction to Navigation

Despite anything you may have heard to the contrary, navigation is a straightforward business. It is built around two simple questions: "Where am I now?" and "How do I safely get from here to where I want to go?"

The most important characteristic which all good navigators share is common sense. You don't need to be a mathematician. Most practical navigation requires nothing beyond an occasional ability to add and subtract simple numbers. There is no obscure science involved at the chart table.

Navigation is essentially a visual affair. The watery world is out there beyond the cockpit for you to look at. In your hand or on the table is a chart which depicts everything significant that you can see, and a lot more that you can't. The traditional instruments you will be employing can be readily understood and used by a child, while the books of tables are virtually self-explanatory once their purpose has been made clear. There are a number of techniques to see you through, and these will be described in the coming chapters. Committing them to memory comes naturally with understanding.

"Most practical navigation requires nothing beyond an occasional ability to add and subtract simple numbers. There is no obscure science involved at the chart table."

The common sense part of piloting lies in determining which areas of your knowledge will serve to answer a specific question as you sail along. You already have the wisdom to decide what is important on a given day. The procedures to help you apply that knowledge are all described in the following pages. You may be amazed to learn that this is all there is to it. Master these orderly concepts, and the rest will come naturally.

The main misconception associated with learning navigation is that piloting is the most important aspect of skippering a yacht. As a result, inexperienced skippers tend to spend too much time at the navigation station and too little out in the cockpit where the real sailing is going on and where dangers can be seen rather than theorized. Sound navigation relies on that vital element of the navigator's make-up: good, plain thinking. Navigators should decide what they need to know, go below and deal with it, then return to where the action is. A yacht is not skippered from the chart table.

Ralph Naranjo photo

Achieving this balance is not only critical, it is also great fun. So don't be "psyched out" by the challenges of navigation. Look forward to it as a satisfying skill that is surprisingly accessible to all straight-thinking people.

Navigation is practiced at a number of levels. Beginners learning to skipper have many things to think about, and navigating is just one of them. Therefore, in an on board sailing course you may find yourself being shown some legitimate short cuts. These might include working from the magnetic compass rather than the true geographic grid from which professionals operate (the two vary slightly). As you progress toward expert yacht navigation you will realize the benefits of "True" plotting and it will become logical for you to work in that format.

This book is primarily intended as a companion volume to a shore-based navigation course where there are no extraneous distractions. The pages which follow work toward making you into the sort of smallcraft pilot who can pick up any boat, anywhere in the world, whatever the conditions, however strong the currents, however large the tides, and operate it with confidence.

Learn all this material thoroughly while you have the chance, then take it afloat and put it into practice. With sound groundwork, you will find that skilled navigation at sea poses no problems at all, and that confidence in where you are and where you are going will free your mind to concentrate on the great pleasures of skippering your yacht.

NAV. NOTES...

"Navigation" and "Piloting" are interchangeable in nautical jargon. They both describe the science of establishing your position and managing your yacht's progress with a minimum of risk and a maximum of enjoyment. In this book, we will be using both words in a context away from the close proximity of land where there is room and time to perform chartwork. Close to shore, among rocks, shoals and increased marine traffic, you must rely more on observation and preparation without constantly plotting your courses and positions. This, we will call "Inshore Pilotage."

Nav. Notes, which appear throughout this book, are intended as supplementary reminders and helpful hints for the reader. Contributions are welcome for future editions of this book.

CHAPTER 2 Charts and Basic Chartwork

Charts form the core of all navigation. They are a two-dimensional representation of a three-dimensional reality. They show, with great accuracy, an area of sea and bordering land, together with any undersea and topographical features important for navigation.

The Globe and Chart Projections

A degree of compromise must be made in order to reproduce the Earth's spherical surface on a flat piece of paper. This is most commonly achieved by what is called a "Mercator Projection." Imagine a light source at the center of a globe projecting the details of its surface onto a chart wrapped like a cylinder around the outside (see left). Notice that the distance between the horizontal lines of latitude is exaggerated toward the poles. This distortion is of little significance to the coastwise navigator, however, who is operating locally and plotting relatively short distances. The major advantage of a Mercator chart is that lines of latitude and longitude form an easy-to-use rectangular grid. Courses from one place to another are drawn as a straight line on the chart.

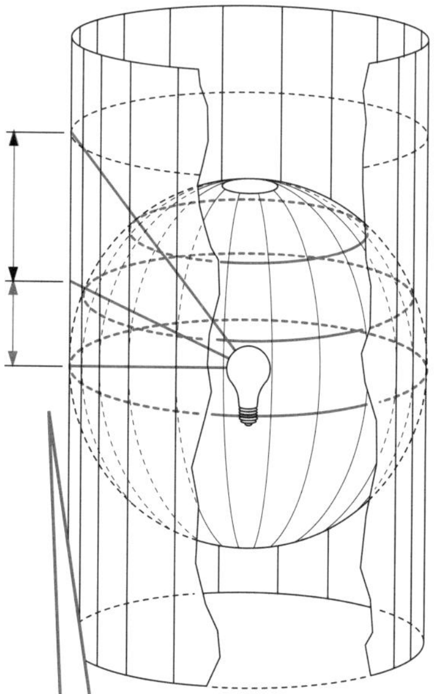

In a Mercator Projection, lines of latitude and longitude form an easy-to-use rectangular grid, but distance between horizontal lines of latitude is exaggerated toward the poles.

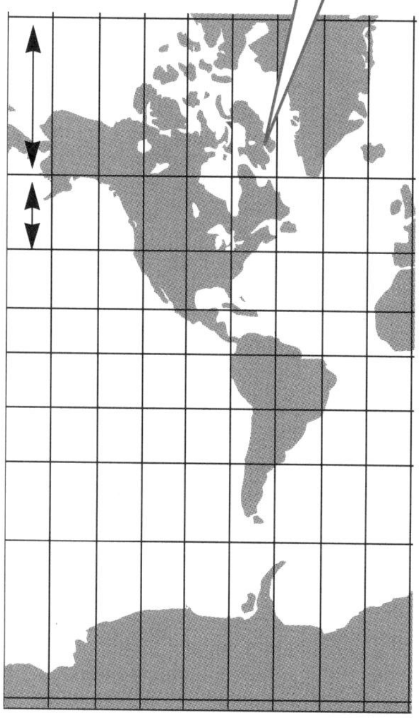

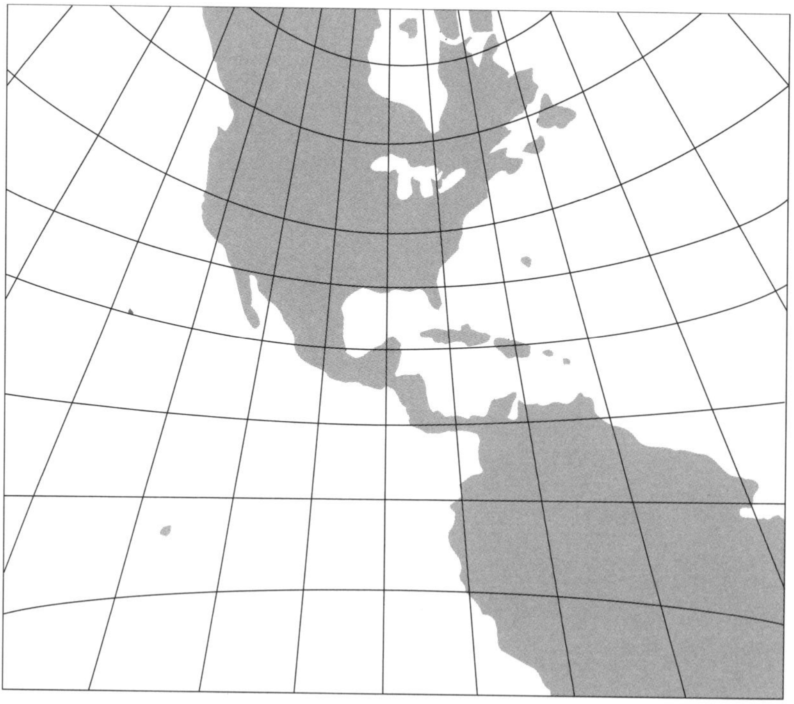

Gnomonic Projection accomodates the curvature of the earth and is often used for long range course planning.

At a large (oceanic) scale, however, the curvature of the earth becomes important and a different kind of projection called Gnomonic is used. On a Gnomonic chart, latitude and longitude lines do not form a parallel grid (*see page 8*), but direct courses from one place to another are still drawn as a straight line. For the most part, coastal navigators use Mercator charts.

The Chart

The coastline is clearly evident on a chart, with sea and land clarified by bold lines and color coding. Everything else is shown in symbolic form, with descriptions often in the form of abbreviations using letters or numbers. The meanings of most symbols should be clear to the non-specialist, but if in doubt, consult NOS (National Ocean Service) chart #1. Note the numbers dotted around the watery parts of the chart. These are *soundings*, or *depths*, and what they mean will be discussed in more detail in Chapter 5. Note also the *contour lines* joining depths that are equal.

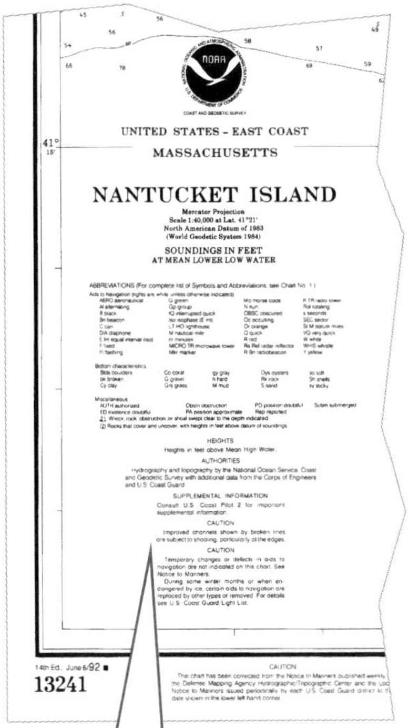

All charts carry a chart title, usually in the least busy corner, and near it is a good deal of important information. The scale is stated here, as is the unit of depth. Units of depth are easy to get wrong if you're not careful because the US is in the process of changing its standard depth unit from fathoms (1 fathom = 6 feet) and feet to the Meter (approximately 3 feet, 3 inches). Make sure you know which unit is being used! In this corner you will also find cautions applying to the area. Check these carefully! Also be sure to check the date of the latest revisions, if any. These are generally noted in the bottom left-hand corner and their presence or lack of it is an indicator as to whether or not the chart is up to date.

Contour lines

Soundings (depths)

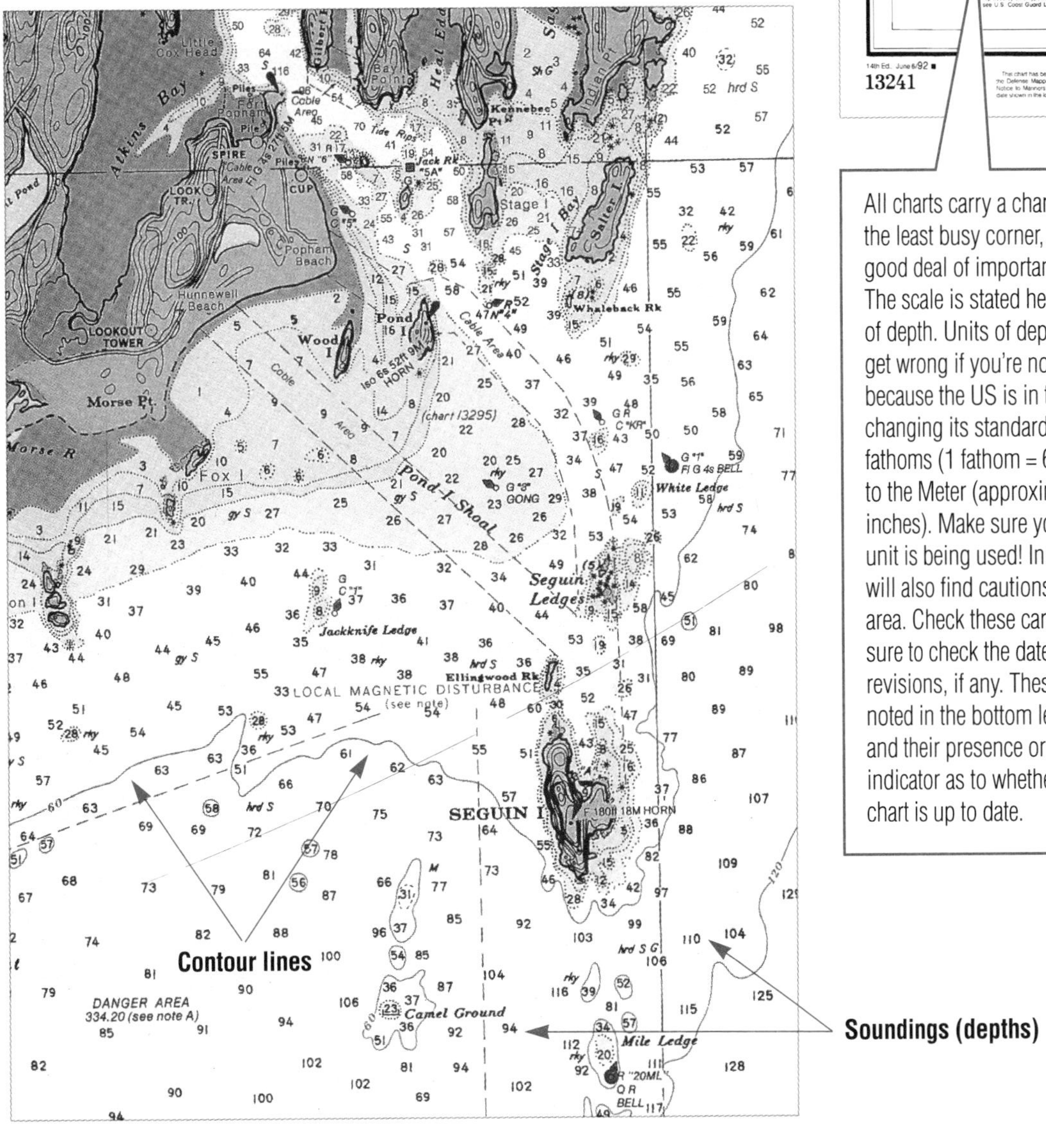

Position

Position on a chart is defined by lines running vertically, known as *longitude* (also called *meridians*), and their horizontal counterparts, called *latitude*.

Latitude is measured in *degrees* (°), 0° to 90° north or south from the Equator, while meridians of longitude are designated from 0° to 180° east or west from Greenwich Observatory in London, England.

A degree is divided into 60 subdivisions known as *minutes* (') – nothing to do with time! – and decimals of a minute. A minute of latitude is equal to one *Nautical Mile* (approximately 2000 yards). Thus, a yacht's position could be given as 37° 40'.6N and 123° 03'.45W (Thirty-seven degrees, forty-point-six minutes North Latitude; one hundred and twenty-three degrees, three-point-four-five minutes West Longitude).

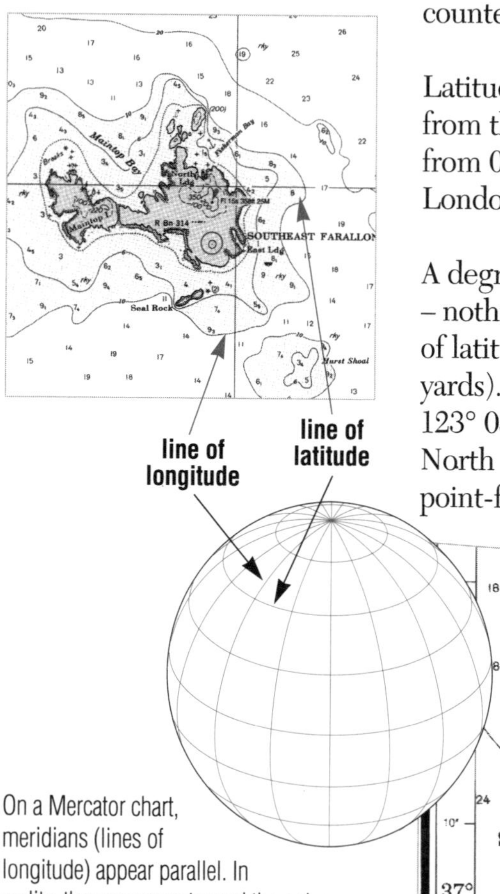

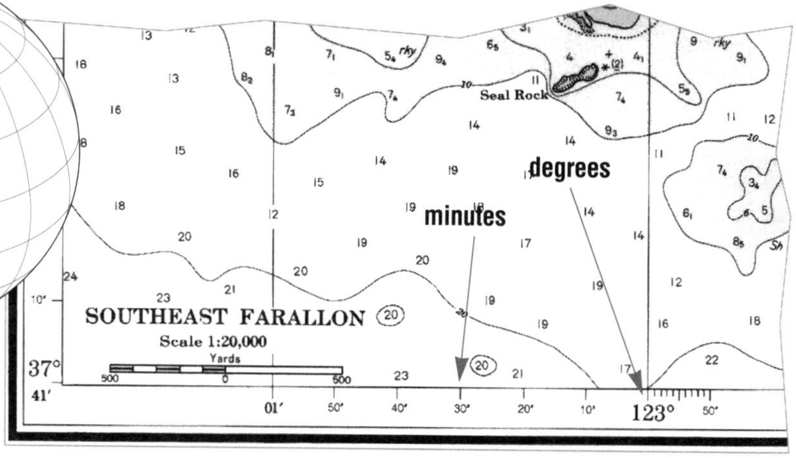

On a Mercator chart, meridians (lines of longitude) appear parallel. In reality, they converge toward the poles (see globe above).

Horizontal lines of latitude run clear around the globe in parallel strips. As a result they are called the *parallels of latitude*. A Mercator chart also shows vertical *lines of longitude* (*meridians*) as parallel, but in reality they are converging, finally meeting at the poles. The amount of this distortion in the lines of longitude varies with latitude. The closer you are to a pole, the more the distortion occurs. On a Mercator chart, minutes of longitude are always of equal size, while minutes of latitude are not of equal size.

None of this is really significant (unless you are planning a trip to Northern Greenland), but it is important to know that if you wish to measure a distance on a chart *you MUST refer to the minutes of LATITUDE*, not longitude, and measure it on the latitude scale parallel to the area where you are working.

Chart Scales

In order to complete a passage of any substance you will require a variety of charts. The first is a *general chart* which, with luck, will accomodate your entire passage. This will be used for planning and for keeping a broad eye on where you are and which way to steer during the main part of the trip. Its scale could be anything from 1:150,000 to 1:600,000. It is unlikely, however, to have enough detail for detailed inshore piloting.

Coast charts are used for entering or leaving bays and harbors. Scales run from 1:50,000 to 1:150,000.

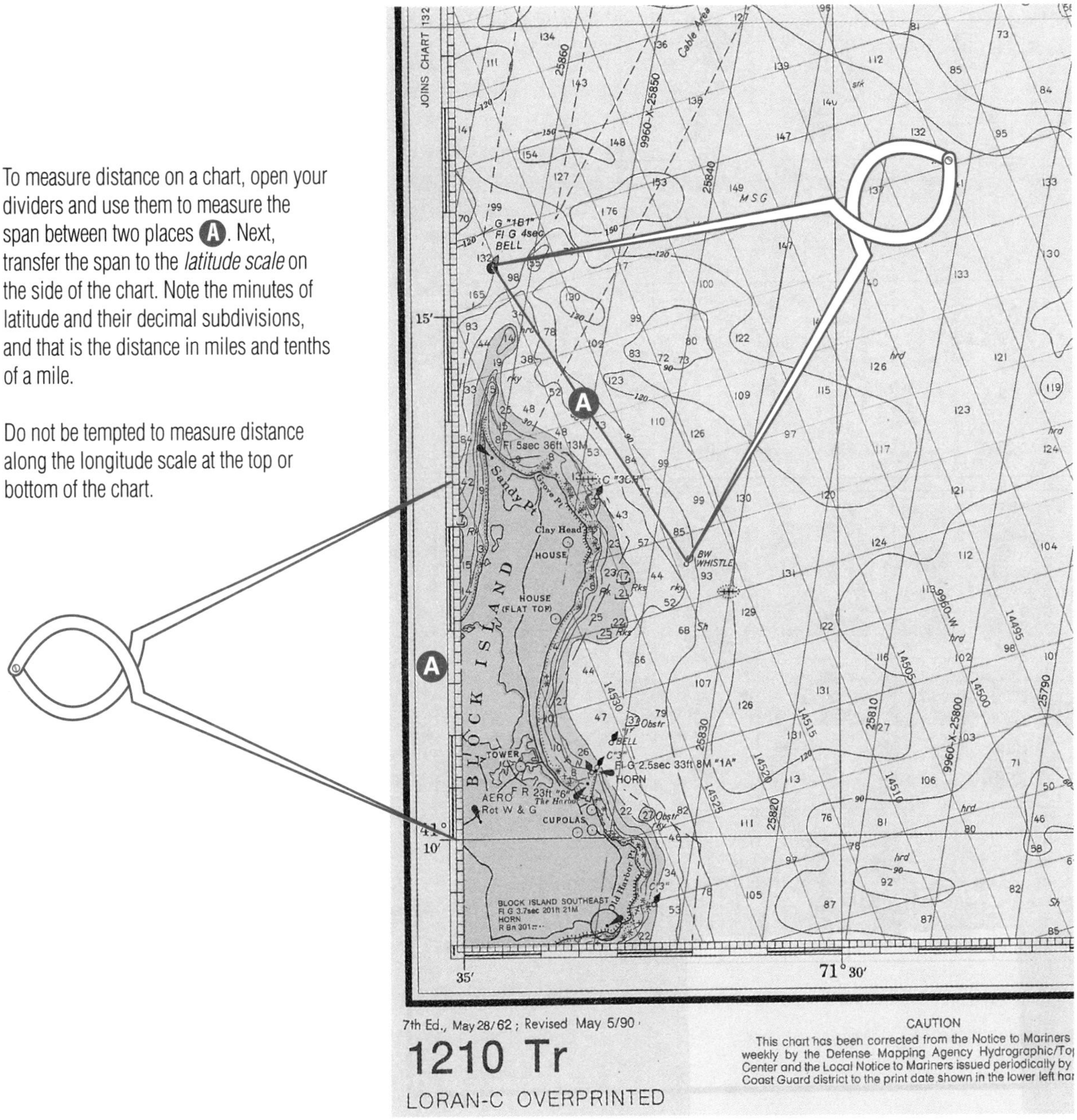

To measure distance on a chart, open your dividers and use them to measure the span between two places **A**. Next, transfer the span to the *latitude scale* on the side of the chart. Note the minutes of latitude and their decimal subdivisions, and that is the distance in miles and tenths of a mile.

Do not be tempted to measure distance along the longitude scale at the top or bottom of the chart.

If there is any degree of complexity about your port of arrival or departure you will also require a *harbor chart* of very large scale. These are sometimes projected gnomonically. This need not bother you, except that when they are, distance scales may be drawn as a distinct item. If you see one of these scales, you should use it rather than minutes of latitude because latitude may be distorted.

Do not worry if you cannot remember the numbers of all these scales. It is sufficient to know that, whatever your requirements, you can tell whether a chart is right for a job by looking at the information it contains.

From time to time you may find that you do not have every chart you need. You should then make the fullest use of Coast Pilots and cruising guides (see Chapter 3), and navigate with the greatest caution, carefully observing buoys if you are in the slightest doubt.

Direction

For a number of reasons it is vital that the navigator can define accurately the direction from one point to another. This is achieved by following the "360° in a circle" principle. While the familiar North, South, East and West naming system still holds good for purposes such as wind direction or remarks about a boat's general heading, any accurate statement must be given in degrees.

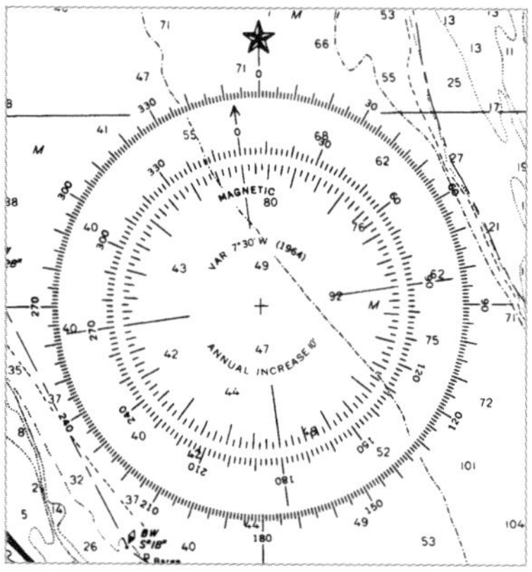

A compass rose is printed in numerous locations on a chart as a reference for plotting courses to steer and establishing your geographic position.

The purest form of directional measurement lines up zero degrees with any meridian of longitude, heading North and South "up and down the chart". To assist in this, all charts carry *compass roses* (see left) which are small circles showing the orientation of the chart and the 360 degrees which make up the circle of possible directions.

Angles on a chart are often measured using an instrument called a *plotter.* This is a form of transparent plastic protractor with a movable arm which lines up with the degrees on its scale. The reading obtained is given in degrees "True" because it uses the chart's latitude and longitude lines as its starting point. Another instrument is a *parallel ruler*, which uses the compass rose as its reference.

Plotting a Course to Steer

A line joining the yacht's position and the point it wants to go to is called a *course to steer*. In the example below, the plotter body is first oriented with the north-south meridians of the chart, then the movable arm is swung around onto the course line. The course here is 060 degrees "True" (T).

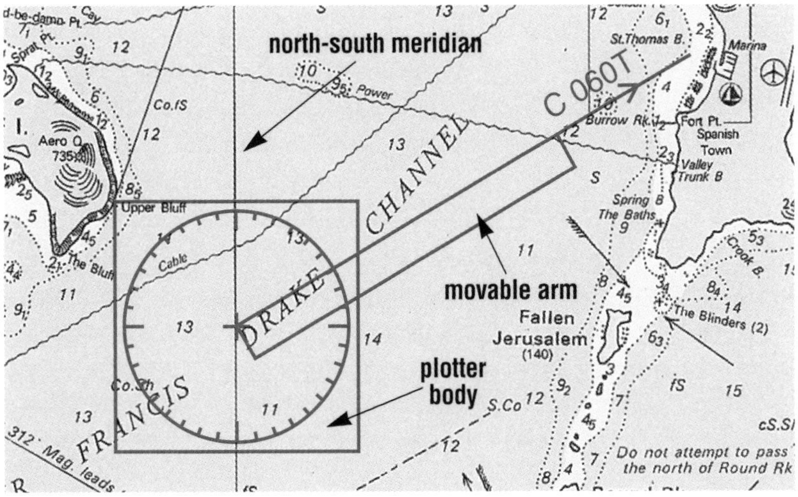

Chuck Place photo

A clear plastic **plotter** measures angles on a chart. It measures in "True" geographic degrees (not magnetic) because it lines up with the geographic latitude and longitude grid on the chart.

Plotting a Bearing

An imaginary line joining a known point with the yacht's position is called a *bearing*. In the illustration below, parallel rulers are used to plot a bearing from a point on land (not a course line) to help establish the yacht's position. Notice the compass rose has an inner and an outer ring. For "true" geographic directions, use the outer ring. The inner, "magnetic" ring will be explained later. In this case, the bearing is 314 degrees "True" (T) on the outer ring and 330 degrees "Magnetic" (M) on the inner ring.

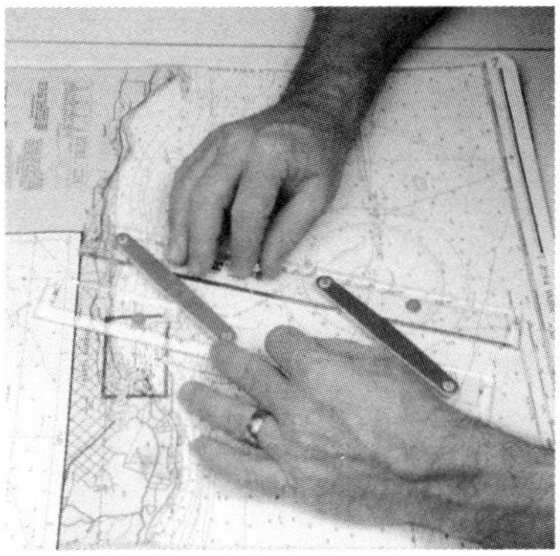

Chuck Place photo

Some navigators prefer to use **parallel rulers** in conjunction with the compass rose printed on the chart to establish direction. Because the ruler sometimes must "walk" across the chart to get from the course to the compass rose, you must be careful that it does not slide out of line.

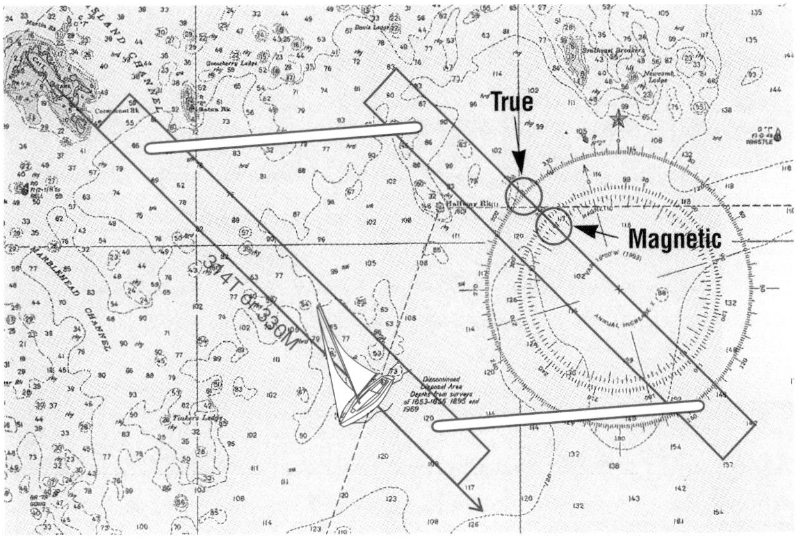

CHAPTER 3 Chart Types & Corrections

Types of Charts

In addition to varying chart scales, you will also discover that there are a number of different chart types which are appropriate to leisure users. This diversity should not be a problem for you, because they all use the same conventions. In theory, certain chart types are preferred for different circumstances but, like most navigators, you will probably end up using what can be found most conveniently.

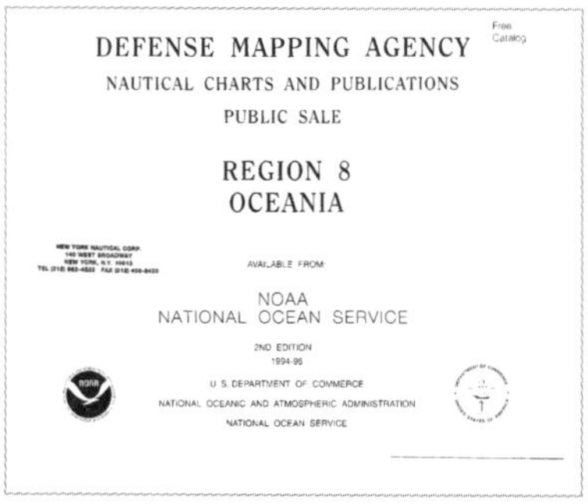

Defense Mapping Agency (DMA) charts are US produced charts of foreign waters.

NOS Charts — These are published by the National Ocean Service. They are large charts which may require a number of folds before they can be accommodated on the average yacht's chart table. The charts are clear, easy to read, and are printed on robust paper.

Smallcraft (SC) Charts — These are charts of inland waters such as the Intracoastal Waterway, and special editions of conventional charts with additional data of interest to leisure sailors, such as tidal predictions, weather bulletin information and shoreside facilities. They are sold folded, and are printed on lighter paper than NOS charts. Such charts may be issued in "book" form, which can be handy on a cruise.

Commercially Produced Charts — In certain areas, charts produced by companies other than NOS may be available that are specifically designed for sailors. IMRAY/IOLAIRE, for example, produce excellent charts of the Caribbean area which feature inset "closeups" of harbors and zones of particular interest. Folded to a handy chart-table size, they come complete with the latest information for up-to-the-minute corrections.

Pilot Books

Because a chart cannot qualify the information it is giving, pilot books are published which supplement charted data with written text, plans and photographs. The books come in the form of official COAST PILOTS and commercially published yachtsmen's pilots, or "cruising guides."

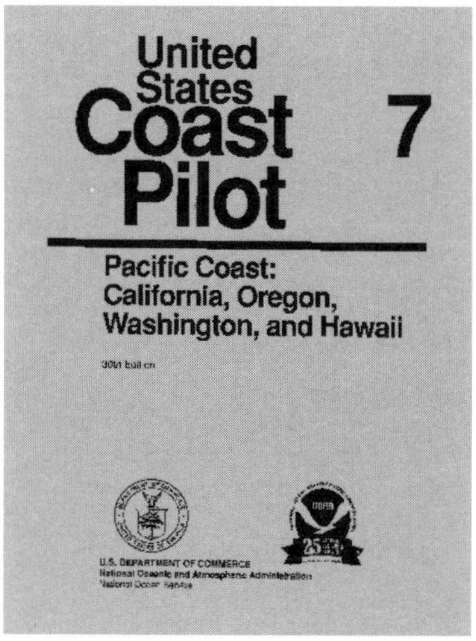

Coast Pilots — Everything technical you need to know is to be found in these books (see left), but they are written from the big ship pilot's point of view. The descriptions of the coastline are of great value, however, as is data on weather, channels and other facets of navigation. Nine volumes of these cover all US waters. If you want official foreign information you will have to go to the SAILING DIRECTIONS.

Cruising Guides fill in the gaps in COAST PILOTS for the local yachtsman, offering valuable knowledge at a modest price. Their information about inshore pilotage can be really useful, and the good ones will even tell whose breakfast to avoid and which marina has a laundromat.

Light Lists are books published regularly with all the latest information concerning lit aids to navigation. If your charts are not up-to-date, a LIGHT LIST (see right) can be a vital safety aid.

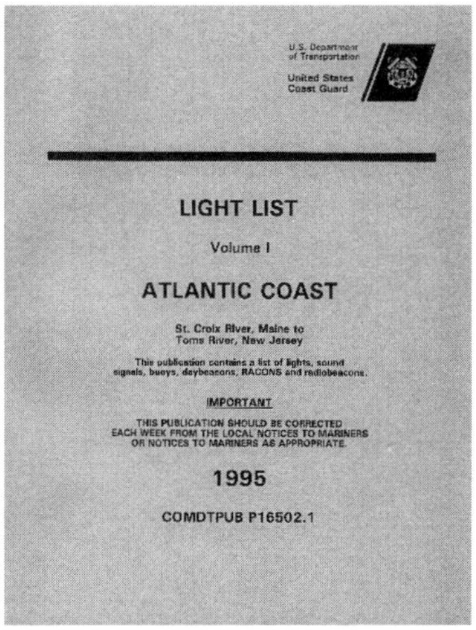

Keeping Charts Up To Date

Because of the changes continually made to navigational aids such as buoys, it is important to keep charts up to date. Every so often a new edition of a much changed chart will be issued, but until it is, you are responsible for doing what you can to stay "on line." NOTICES TO MARINERS are issued weekly, listing all changes world-wide. No recreational sailor needs more than a hundredth of this information, so the most appropriate way to maintain your charts is to be aware of LOCAL NOTICES TO MARINERS. These are also published weekly, but only for your area, by the District US Coast Guard. It is a generally accepted convention to note down all corrections made with their year and notice number in the bottom left-hand corner of the chart.

Living with Uncorrected Charts

Most private sailors do not correct their charts on a regular, disciplined basis because it can be time consuming. If a chart is known or suspected to be out of date, first check its date of issue. Next, note any buoys or lights you are going to be interested in and check them off against your latest LIGHT LIST or almanac. Rocks don't usually move, but be ready for changes in less solid bottoms. Sand bars can creep and coral grows actively in some areas.

Most important of all, never assume anything when working from an old chart. If you try to match what you see with a chart that doesn't show it, you could end up confused and in trouble. Caution at all times is the watchword.

CHAPTER 4 Aids to Navigation

For more than a thousand years, mankind has placed piles of stones in the sea or anchored floating barrels to the seabed to indicate the whereabouts of nearby dangers, above and below the water. Today, we have lighthouses that can be identified up to 30 miles away, and buoys whose lights, colors and shapes give mariners detailed guidance, especially in unfamiliar waters.

All significant aids to navigation are charted, so when you see one on the water, your first job is to identify it. Conversely, you will often find yourself looking for a specific navigation aid which is shown on your chart. Read this chapter carefully!

Ralph Naranjo photo

Lighthouses

These are powerful lights that are often, though not always, visible far out to sea. A complete reference on lighthouses is provided by the Light List, but your chart and your pilot books will tell you most of what you need to know.

Light signals — Lighthouses identify themselves by characteristic flashing sequences. A flashing light emits its signal in short bursts with periods of darkness between. An occulting light is on most of the time and periodically winks off. Sometimes called a "black flash," occulting lights are easiest to take bearings on.

Most charts indicate a lighthouse with a magenta symbol that looks like an exclamation point (see example below). Information about the light type and sequence are indicated by various codes. (Full details are in NOAA Chart #1).

This chart shows the lantern of Southeast Farralon Light standing 358 feet above mean high water. It flashes once every 15 seconds and its range of visibility is 25 miles.

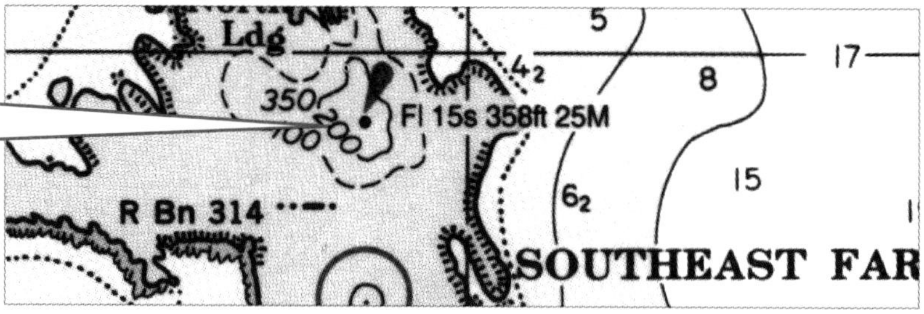

Another example might be: Fl(2) 5s 10m 11M = Group Flash 2 every 5 seconds. Height of lantern, 10 meters. Nominal range of light, 11 miles. "Group Flash" means that the light winks several times in succession, followed by a period of darkness. Notice that "Miles" is expressed by a capital "M", while "meters" takes a small "m". It is important to note that on metric charts, the heights of lights and other items are given in meters, not feet.

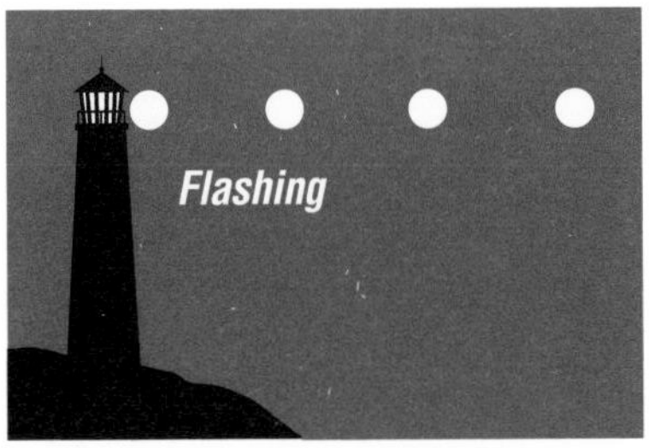

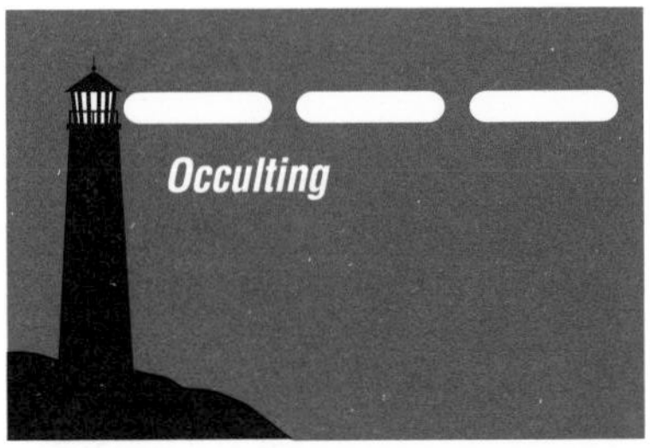

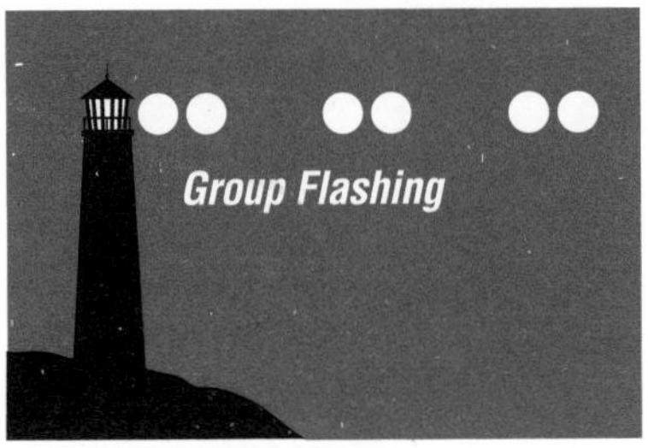

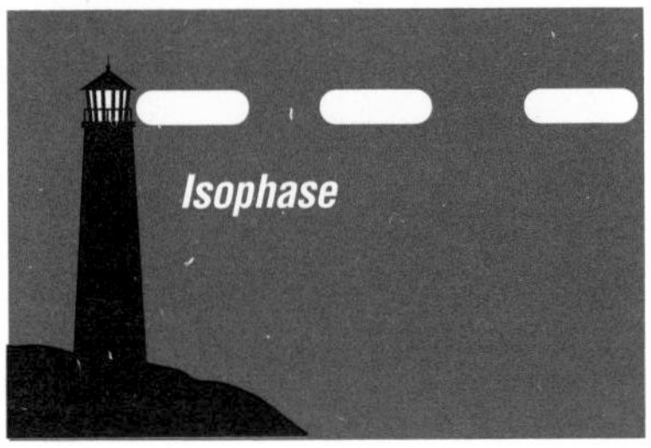

A **flashing** light is on in short bursts interspersed with longer periods of darkness. A **group flashing** light repeats multiple light signals. A **composite flashing** light repeats irregular multiples of signals, or example "2+1". This means that within its time cycle the light will flash a group of 2 flashes followed by a short pause, then a third flash.

An **occulting** light is on most of the time and "winks" off according to its charted sequence. Sometimes called a "black flash," these lights are easy to take bearings on. An **isophase** light has equal periods of light and darkness.

Range of Visibility — This is defined as the maximum distance a light can be seen in clear weather. The range of visibility is given on the chart and is called the Nominal Range. The same range is given in the Light List publication.

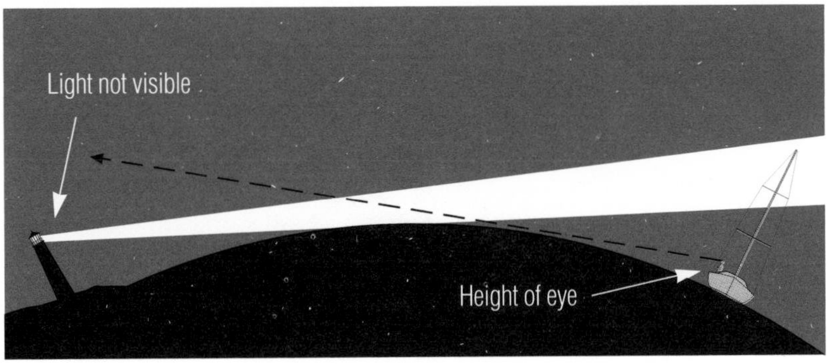

Whether you can actually see a light or not will depend on the clarity of the air and your height of eye (how high you are standing above the sea). If a light as bright as the sun were displayed from a height of 10 feet, you would not see it from the deck of your boat fifteen miles away. What you would see is the loom, a diffused glow from the light in the sky.

Height Feet	Nautical miles	Statute miles	Height *meters*	Height Feet	Nautical miles	Statute miles	Height *meters*
1	1.2	1.3	0.30	120	12.8	14.7	36.58
2	1.7	1.9	0.61	125	13.1	15.1	38.10
3	2.0	2.3	0.91	130	13.3	15.4	39.62
4	2.3	2.7	1.22	135	13.6	15.6	41.15
5	2.6	3.0	1.52	140	13.8	15.9	42.67
6	2.9	3.3	1.83	145	14.1	16.2	44.20
7	3.1	3.6	2.13	150	14.3	16.5	45.72
8	3.3	3.8	2.44	160	14.8	17.0	48.77
9	3.5	4.0	2.74	170	15.3	17.6	51.82
10	3.7	4.3	3.05	180	15.7	18.1	54.86
11	3.9	4.5	3.35	190	16.1	18.6	57.91
12	4.1	4.7	3.66	200	16.5	19.0	60.96
13	4.2	4.9	3.96	210	17.0	19.5	64.01
14	4.4	5.0	4.27	220	17.4	20.0	67.06
15	4.5	5.2	4.57	230	17.7	20.4	70.10
16	4.7	5.4	4.88	240	18.1	20.9	73.15
17	4.8	5.6	5.18	250	18.5	21.3	76.20
18	5.0	5.7	5.49	260	18.9	21.7	79.25
19	5.1	5.9	5.79	270	19.2	22.1	82.30
20	5.2	6.0	6.10	[illegible]	[illegible]	22.5	85.34

This Distance of Visibility of Objects at Sea chart gives the approximate range of visibility of an object which may be seen by an observer at sea level. By selecting both the height of the object and the height of the observer, then adding their corresponding visible ranges, you can determine the visible range for any particular situation. **Example:**

Height of object: 18 ft. 5.0 mi
Height of observer: 9 ft. +3.5 mi
Geographic visibility: 8.5 mi

Lighthouses are often conspicuous structures which can be readily spotted from quite long range by day, especially if you use your binoculars. For descriptions, see the Light List, or your pilot book.

Ralph Naranjo photos

Sound Signals

Many light structures and buoys emit sound signals in poor visibility and fog. These can be made by various types of equipment and sound as follows:

Diaphone (DIA)	Grunting sound
Horn (HORN)	Various steady pitches. Horns are often the modern Nautophone device, which generates a high, extremely penetrating note.
Siren (SIREN)	Generally higher pitched than a horn or a diaphone
Whistles (WHIS)	Often found on buoys — various pitches
Bell (BELL)	Self-explanatory. Usually on buoys
Gong (GONG)	Combination of several different and distinct gong pitches

(Some lighthouses also emit an explosive sound)

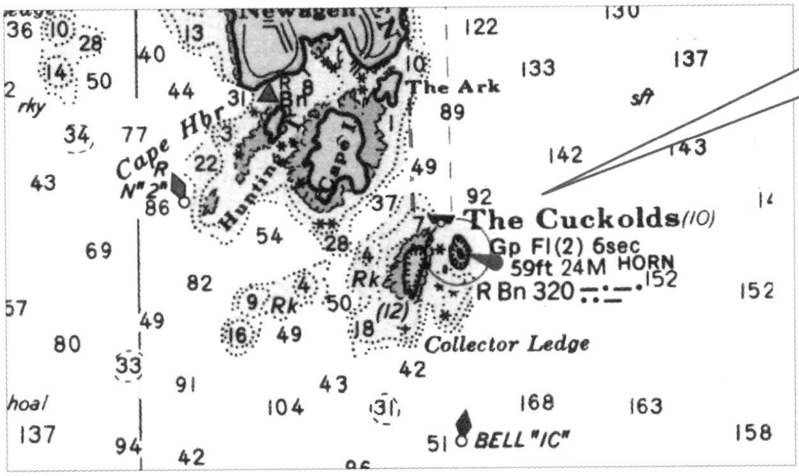

This lighthouse shows a light which "group flashes" 2 times every 6 seconds. Its sound signal is a horn.

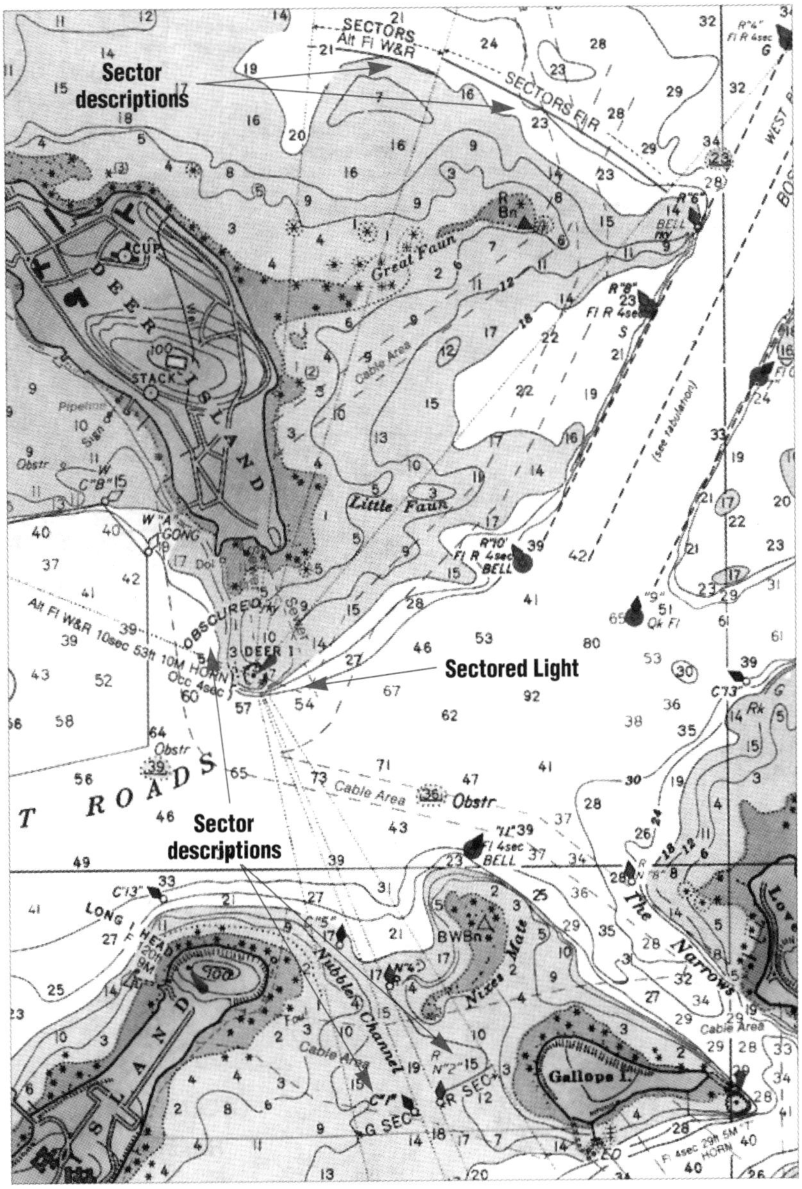

Some lights are sectored in different colors or intensities to indicate the presence of channels or dangers, but will not have different lights visible from different heights. The sectors are marked on the chart and are given in the Light Lists expressed in numbers of degrees.

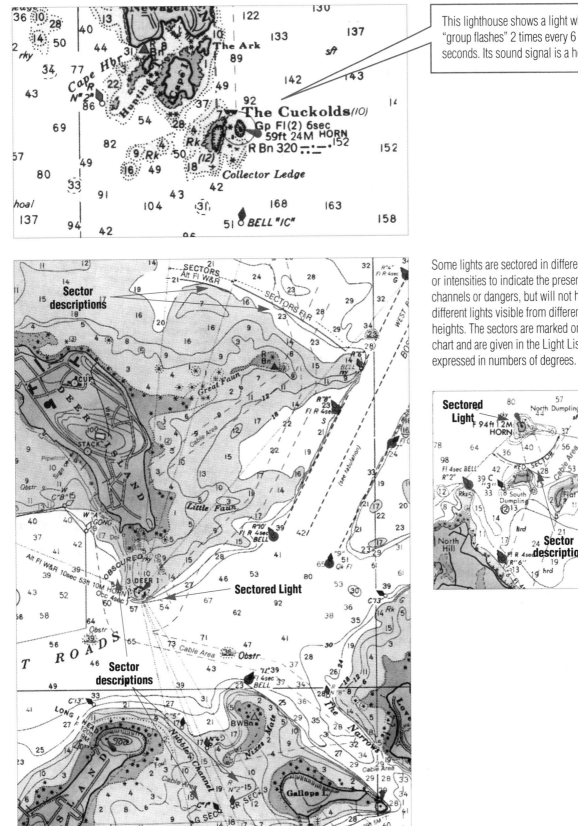

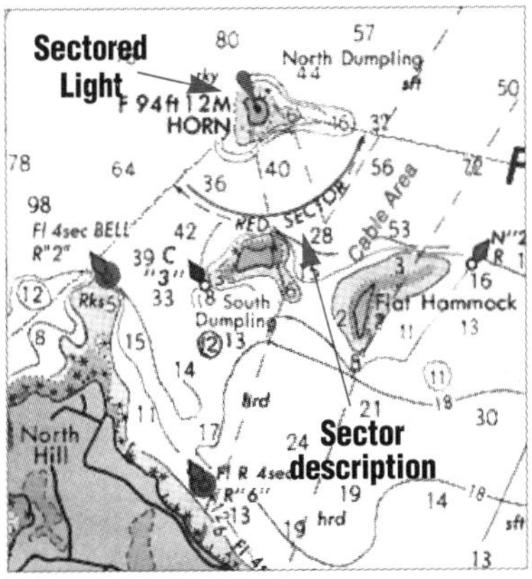

Buoyage

Buoys are placed to define deep water channels and to indicate places of danger. They can be small or surprisingly large, lit or unlit. At first glance, their systems of identification appear complex but they are in fact remarkably logical.

Lateral marks are placed on either side of a channel, green, flat-topped "cans" to port and red, pointed-topped "nuns" to starboard. Large channel buoys may be of steel lattice construction and generally cylindrical in shape. In such cases, they are often equipped with *topmarks* depicting the flat-topped can or the pointed nun form.

Cans have flat tops and odd numbers

Nuns have pointed tops and even numbers

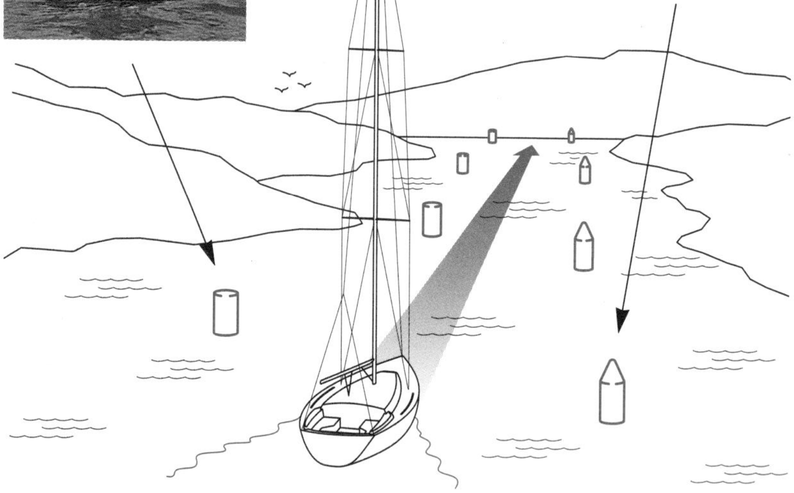

When approaching a channel from seaward, keep green cans to port and red nuns to starboard. Remember the phrase..."RED RIGHT RETURNING." The black port-hand cans of an older system are being replaced steadily with greens. Sometimes, you will see smaller "spar" buoys instead of cans or nuns. These are colored according to their lateral meaning.

Note that red buoys on the chart have an "R" showing their color, while greens often do not. Note also that greens number odd (1, 3, 5, etc.) from seaward, while reds number even. If the nun or can has a light and/or sound signal, this is also noted on the chart.

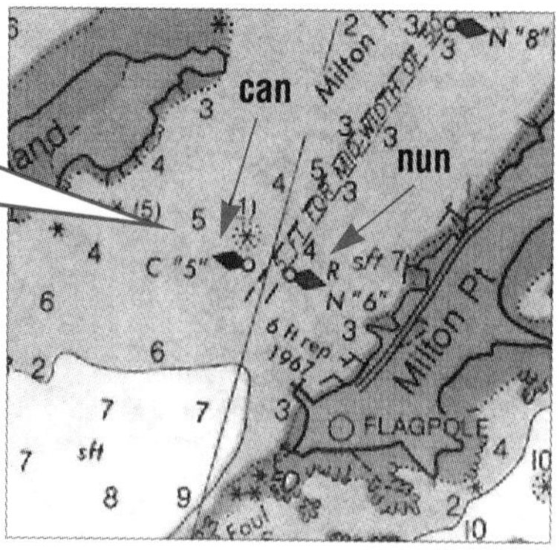

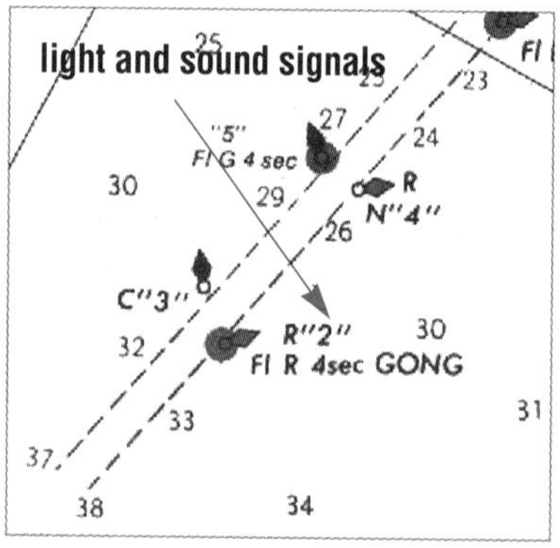

Lighted Buoys — Red buoys, if lit, carry red lights while greens show green. Light sequences are indicated on the chart.

Just because a buoy exists does not necessarily mean you must obey it. Refer to your chart to check its relevance. It might only be important to big ships, in which case you could be better off passing the "wrong" side of it in order to keep well clear of commercial traffic. Always read your chart and make a sensible judgment. Special buoys other than for direct navigation can be any shape and may be various colors. When you see these, refer to the chart to check if they have any significance to you. If in doubt, keep clear to seaward.

Do your best to learn these buoy types, but don't break your heart if you have trouble cramming them all at once. In practice, so long as you can observe a buoy's color, topmark and its light or sound, you can usually find it on the chart, especially if it has a number! Binoculars are an enormous help in real-life buoy recognition.

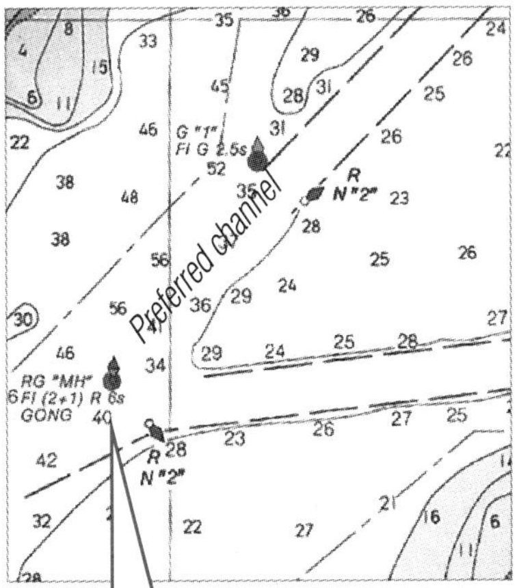

Where channels divide, one will be "preferred" for deeper draft vessels. It may or may not be the one you want, but at the division you will find a **preferred channel buoy**. If the main channel is to your left, when coming from seaward, the buoy will be left to starboard by main-channel users and it will be colored in horizontal red and green bands. Logically enough, the red band is uppermost. Where the preferred channel lies to your right, the green band is at the top because the buoy is to be left to port by deep-draft shipping.

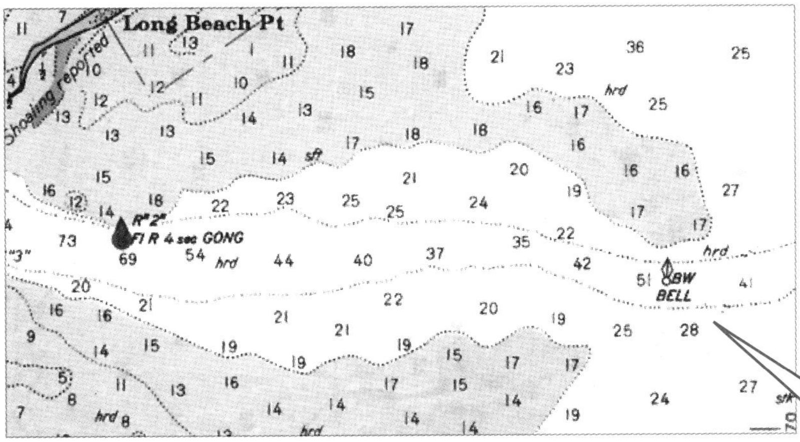

Sometimes you will find a **safe water mark** when approaching a channel from seaward. These are colloquially known as **"fairway buoys."** They are red and white vertically striped, often spherical in form. If lit, they flash a Morse code "A" (• —) about 8 times per minute. These buoys are never numbered, but are sometimes lettered.

Wrecks are generally marked with green buoys which will be placed as close to the wreck in a seaward direction (or toward the channel) as possible. Be careful! The sea can shift a wreck, so don't go too close. Occasionally you will see wrecks marked with buoys painted in green/red or red/green configuration. These refer to "preferred side to pass." In waters where the cardinal system is used, wrecks may be surrounded by four cardinal buoys.

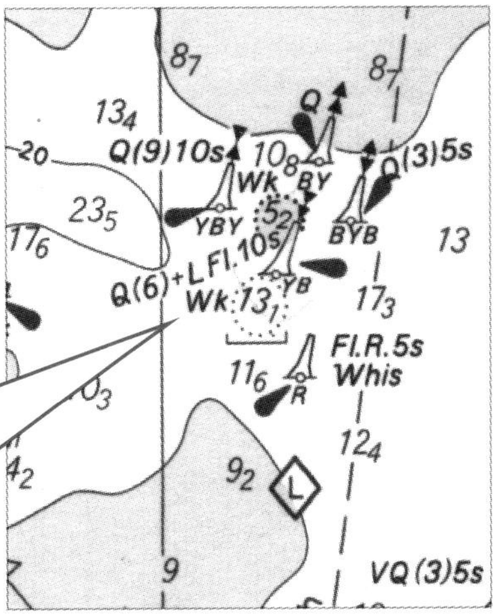

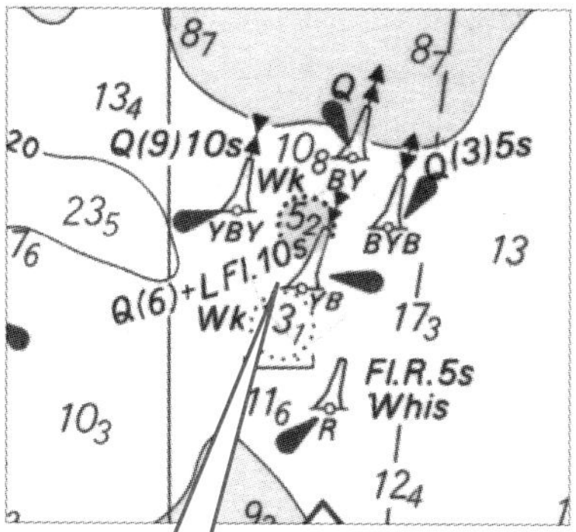

Isolated dangers may be marked with red and black horizontal striped buoys or beacons (lateral and cardinal systems).

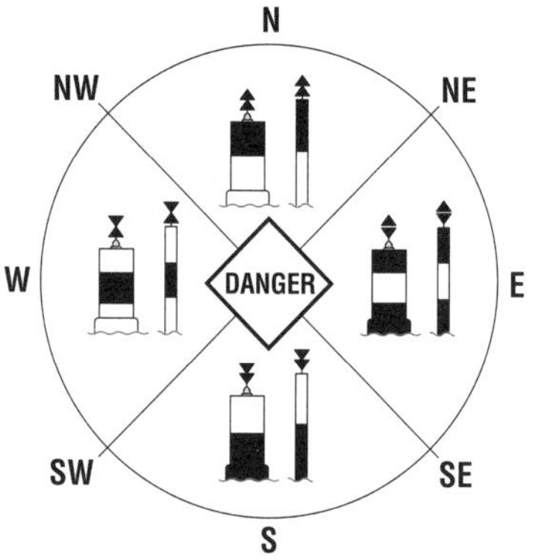

Cardinal marks, are uncommon in US waters. They designate a safe side on which to pass a danger or the side on which a change occurs. They feature black and yellow bands topped with black cones or triangles that indicate direction.

North	2 triangles point up
South	2 triangles point down
East	triangles form a diamond, and "diamonds come from the east"
West	2 triangles form a sideways W

Unlike buoys, which swing around on moorings, beacons (shown at right) and daymarks are mounted on permanent supports driven into the bottom. They provide precise reference and are especially well suited to tight quarters.

Cardinal Marks

Uncommon in many U.S. cruising areas, cardinal marks designate the side where there is deepest water, the side on which to pass a danger, or the side on which some change occurs. Fixed or floating, the mark features black and yellow bands topped with two black cones or triangles that indicate direction. At night, white lights flash a designated sequence.

Beacons

Beacons are used in some circumstances in preference to buoys to mark almost anything a buoy might be used for. They are built up from the seabed, or may be a stake driven into it. Their colors coincide with buoys, as do their lights. They usually show a topmark to indicate whether they are "nuns" or "cans."

You will also come across daymarks indicating the position of a channel. Once again, their shapes echo the corresponding buoys. Needless to say, these are unlit.

Beacons and daymarks have the advantage over buoys in that they do not surge around on their moorings. A buoy must inevitably have at least some radius of drift. This makes beacons particularly viable in tight quarters, or where the position of a navigational aid must be absolutely exact.

By Night — Nearly all marks are equipped with reflective tape for your searchlight if you are hunting for an unlit one in the dark. Be careful with your light. Don't blind your crew with it, and be mindful of other water users, particularly those coming towards you down the same channel.

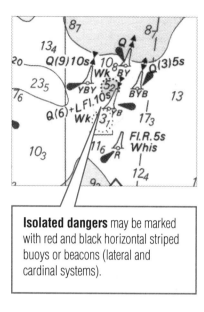

CHAPTER 5 Navigational Inputs

Various simple instruments, including your own eyes (the best of all), are used to transfer information from the real world to the small flat one that is the chart. Here are the principal ones.

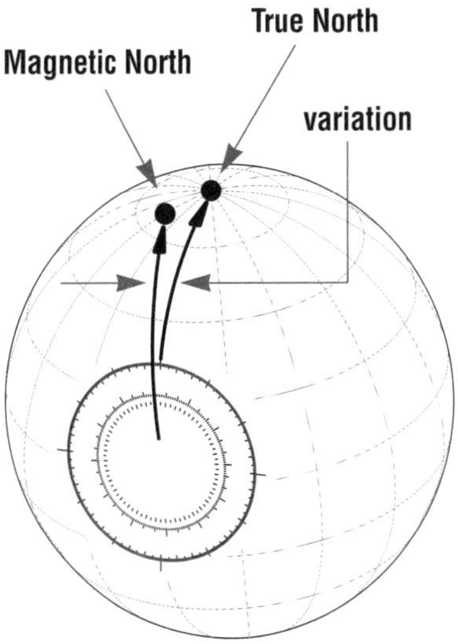

The amount and direction by which the Magnetic North varies from True North is noted on a chart and is called **variation.** You will find the numbers inside the compass rose (see below). This example indicates a variation of 7°30' West (W). The variation is increasing by 10' (ten minutes of arc, or one sixth of a degree) annually. Look at the date on the chart. If it is already 6 years old, the variation will have increased by 10 minutes x 6 = 60 minutes, or 1 degree, leaving a current figure of 8°30' W.

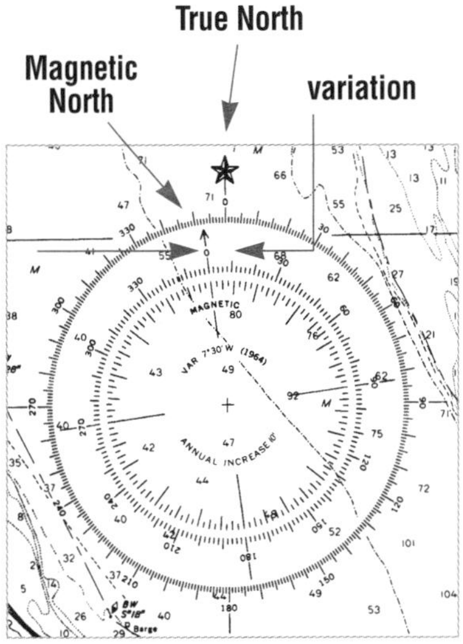

The Eyeball

The first and possibly the most important navigation lesson you can learn is to stop your boat in a known position (near to a buoy, for example), pick up a chart, orient it north / south / east / west, and look around you. How much of your visible surroundings can you relate to what the chart depicts? Take some time, refer to Chart #1 if you need to clarify the chart symbols, and sort out "what is what." When you come to navigate for real, your eyes are your primary source of data. *Never let theory detract from what you can see!*

In reality, much of your inshore piloting will be done on eyeball alone. Experienced navigators, however, also know the danger of being too casual. If you are in doubt, always plot a fix and a course to steer. If you are confident, and that confidence is built on sound knowledge and good reason, then don't feel you must always plot on the chart.

The Compass

Variation — A yacht's steering compass lines itself up with the Earth's magnetic field. The "magnetic" North pole, however, is different from the "true" geographic North pole. In reality, "magnetic North" slowly wanders in the Canadian Arctic. Your compass will therefore point to "magnetic North" and all its bearings and courses will be magnetic. To ensure that there is never any confusion over how a course is designated, magnetic headings or bearings should be given the suffix, "M," while true geographic courses are denoted "T."

If you are navigating using degrees True (T), which is a sound plan, you must adjust your courses to steer to "Magnetic" before passing them to the helm, and any magnetic bearings handed down to you from the helm must be converted to True.

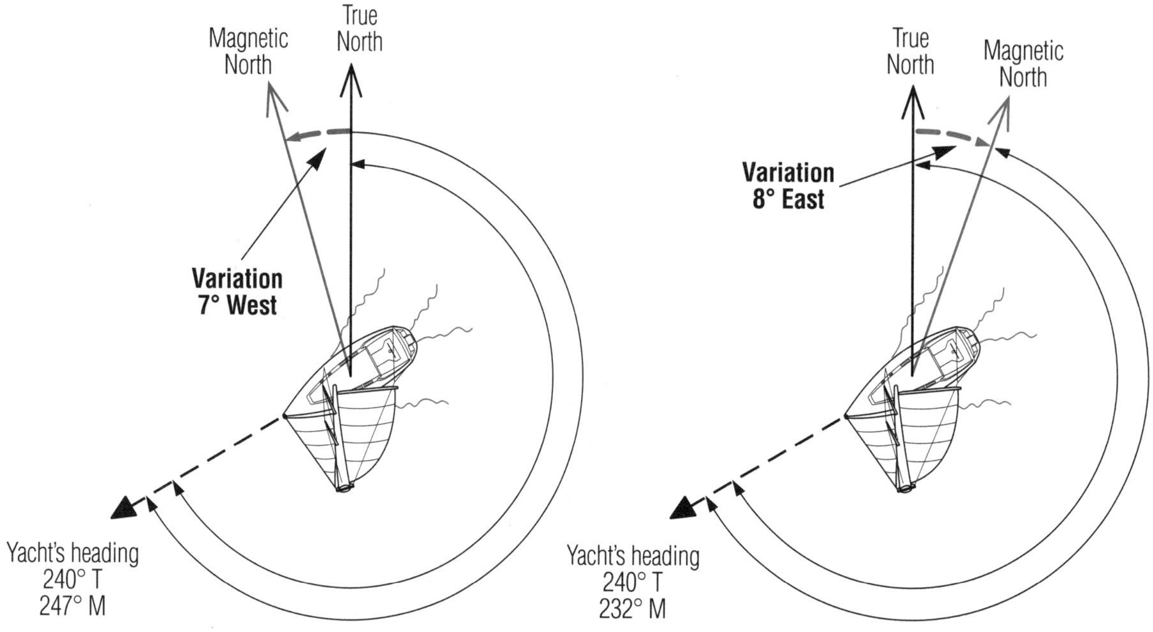

If in doubt about whether to add or subtract variation, always draw a sketch as shown. This makes the answer obvious, because the line in the left illustration, from the ship's head to the magnetic north sweeps through more degrees than the one joining it to the true north. The magnetic heading is therefore greater with Westerly variation. Vice versa with the right illustration, where variation is Easterly.

From these diagrams you will see that if the variation is West, the compass will give a bigger number than the True reading. If variation is East, the compass will have a smaller number. This is taken care of by a simple mnemonic...

Error West, Compass Best (Biggest)
Error East, Compass Least

This suits some people but not all. If you cannot get on with the mnemonic, stick to the sketches.

Deviation — Your ship's compass may also be subject to deviation (error) resulting from magnetically active items on board. Many modern yachts are largely unaffected by this, but for some, deviation (which varies with the yacht's heading) is a significant factor. A professional compass adjuster can tabulate these compass deviations in a *deviation card* which should be kept in the chart area (*see page 28*). Deviation may run up to 10° for a steel or ferro-cement yacht. Deviation must also be factored when you convert to and from True headings. If no deviation card is present where one is needed, disappointing landfalls will result. Once a season you should check your own compass by *swinging ship* (*see page 27*).

If a compass has deviation, a third type of heading is introduced in addition to "True" and "Magnetic." A heading taking into account *both* deviation and variation is called a *compass heading* and is designated by the letter "C". If there is no deviation, you could call a magnetic heading a compass heading but this is a poor plan, because it leads to sloppy thinking.

In order to convert from a True to a Compass heading, first apply variation to arrive at "Magnetic." Next apply deviation to the magnetic value to find the compass heading.

True Heading	**T**rue	**T**ele-
↕	↕	↕
apply **Variation** to give	**V**ariation	**V**isions
↕	↕	↕
Magnetic Heading	**M**agnetic	**M**ake
↕	↕	↕
apply **Deviation** to give	**D**eviation	**D**ull
↕	↕	↕
Compass Heading	**C**ompass	**C**hildren

For generations, a "salty sailor" mnemonic (True Virgins Make Dull Companions) has been used to take compass bearings or headings from True to Compass and back again. In these modern times, however, a more politically correct mnemonic (right) has taken its place.

Deviation can be caused by any heavy ferrous item such as an engine, but it might also be helped along by much lighter objects closer to the compass. When this boat turns, the compass needle should stay still as the boat swings beneath it, but it is pulled from its proper bearing by the engine. Because the relative position of the engine and the compass needle changes with the boat's heading, it is necessary to plot a deviation card (see page 28) showing the change in deviation with eight or sixteen headings around the compass.

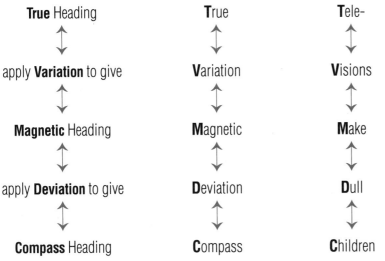

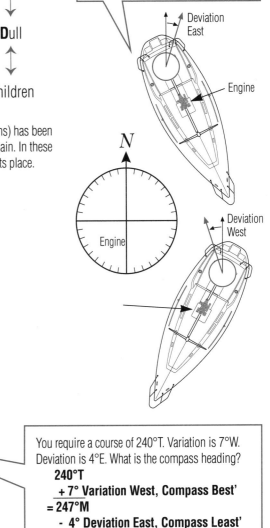

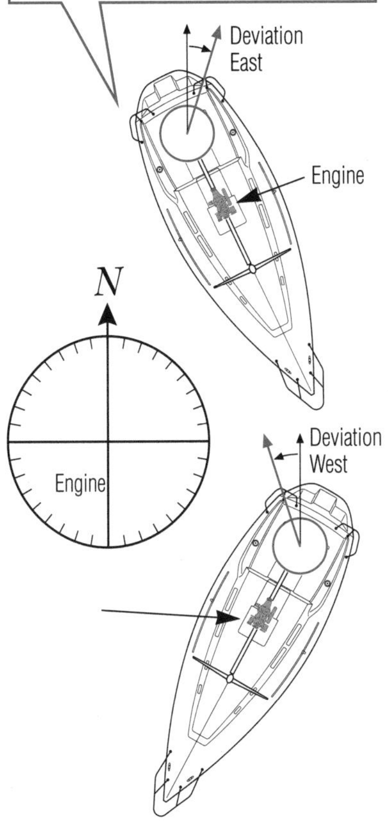

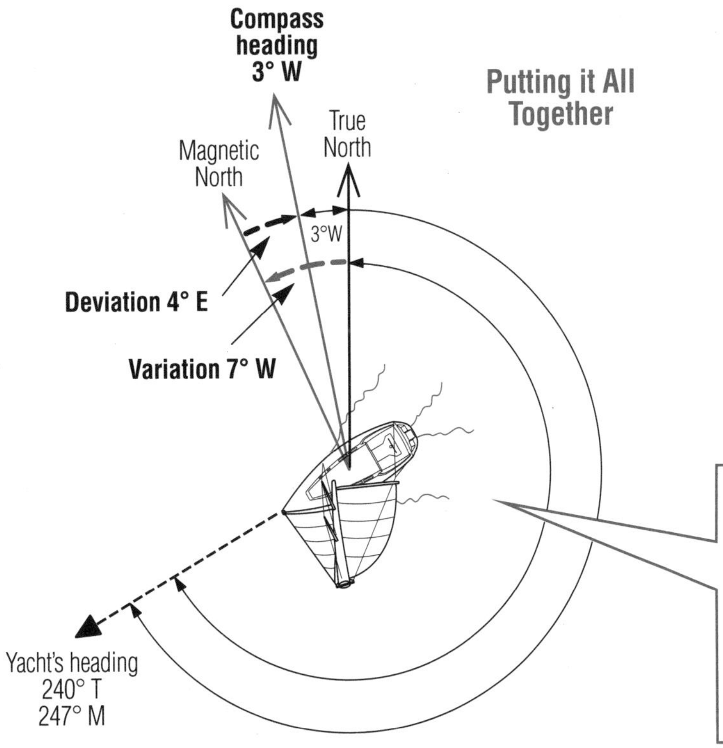

You require a course of 240°T. Variation is 7°W. Deviation is 4°E. What is the compass heading?

240°T
+ 7° Variation West, Compass Best'
= 247°M
- 4° Deviation East, Compass Least'
= 243° C

Do not be dismayed by this business of compass error. In practice, it creates few difficulties because you will usually be sailing in a localized area where variation will remain more or less constant. Once you have applied variation a few times, the technique will become second nature. From there, dealing with deviation (if it exists) is another short step.

Remember that if a compass has no deviation, all courses will be in "magnetic," and are given the label "M". "C" is reserved for courses or bearings where deviation has been applied.

Plotting in "Magnetic" — If your compass has been proved by swinging ship to have no significant deviation, there is nothing to stop you from doing most of your plotting in magnetic. You can use the magnetic compass rose together with parallel rulers, and stick with that denomination.

Be careful if you opt for this policy. Remember to label all plotted bearings and headings "M", because many bearings or ranges spelled out on the chart or in a Coast Pilot will be given in degrees "True." So will current vectors. Deviation aside, this is another reason why the Navy and the Merchant fleet always plot in "True." If you opt for the apparently simpler "magnetic" plotting, never forget that deep down, True is right.

Heeling Error

Just as deviation may be caused by metal objects onboard affecting the compass as the yacht's heading varies, so the relative positions of the same objects can alter significantly at high angles of heel. This sometimes happens where there is a heavy iron keel. It is generally impractical to calibrate this "heeling error," and it is usually small in any case. If you suspect you may be experiencing some heeling error on a particular leg of a passage, check your steering compass against your hand bearing compass.

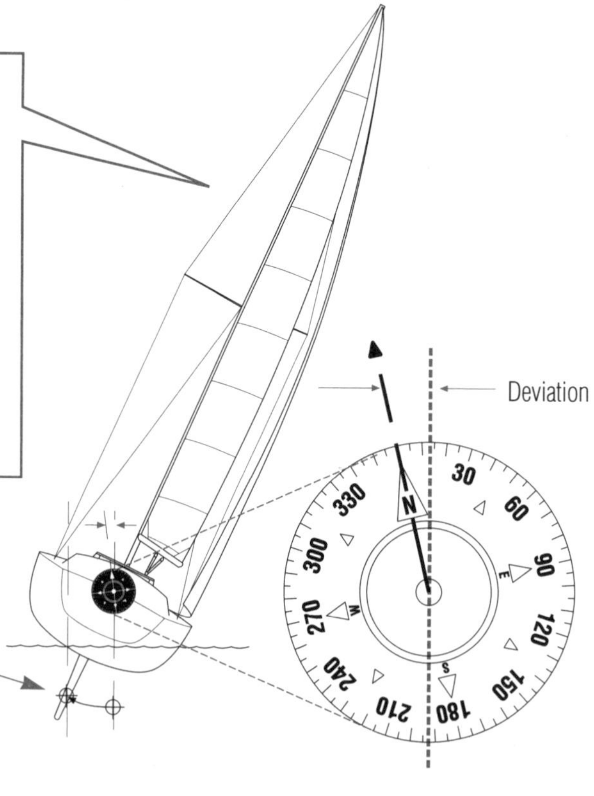

Swinging Ship — Compass deviation is far more common than you might think, and can have a significant effect on the accuracy of your efforts. *Unknown compass characteristics = bad landfalls!* It is as simple as that.

Many boats experience no deviation, but you owe it to yourself to make sure. In order to discover and tabulate any deviation you may have, it is necessary to *swing ship*. This process takes about 20 minutes and can give very accurate results.

To swing ship, stand on deck with a hand bearing compass and sight on any conspicuous object at least three miles away (*see right*). The object need not be charted. Note its bearing and continue to watch it as your mate steers the boat slowly round a circle of, say, fifty yards diameter. Fifty yards will not be enough to materially alter the bearing of an object three miles away, so any change in the bearing must be due to deviation of your hand bearing compass at that spot on deck. Try it somewhere else until you are successful. Knowing where the hand bearing compass is deviation-free is also useful for taking bearings and for checking your steering compass on a passage.

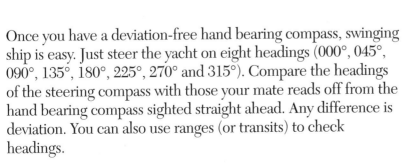

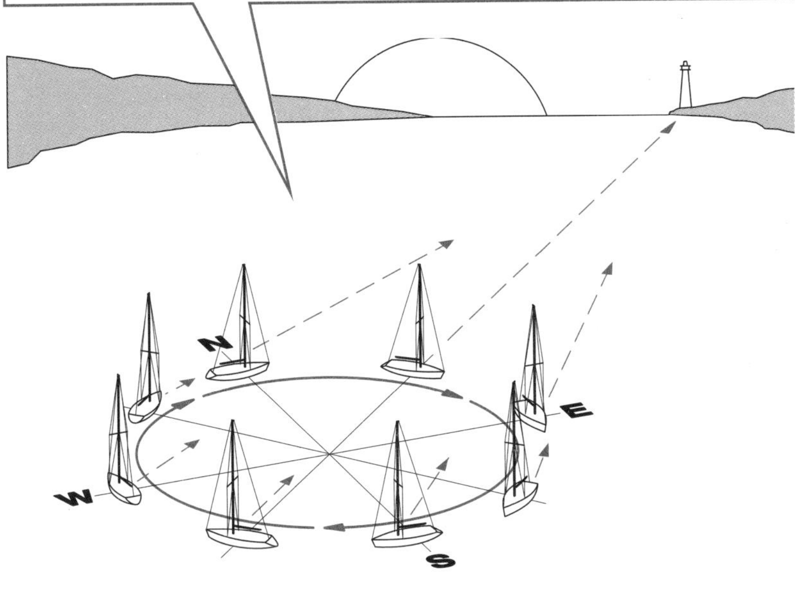

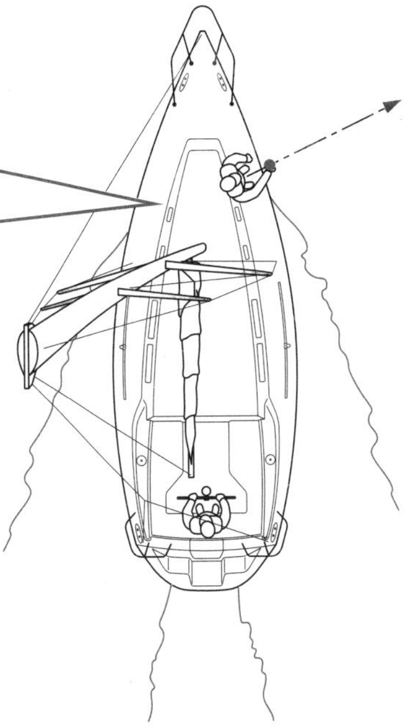

Before swinging ship, you'll need to find a point on deck where your hand bearing compass (the free compass which is not attached to the ship) will have no deviation. Hand bearing compasses are often assumed accurate anywhere onboard, but in reality they can also be affected by all sorts of unlikely items, particularly in the cockpit.

Once you have a deviation-free hand bearing compass, swinging ship is easy. Just steer the yacht on eight headings (000°, 045°, 090°, 135°, 180°, 225°, 270° and 315°). Compare the headings of the steering compass with those your mate reads off from the hand bearing compass sighted straight ahead. Any difference is deviation. You can also use ranges (or transits) to check headings.

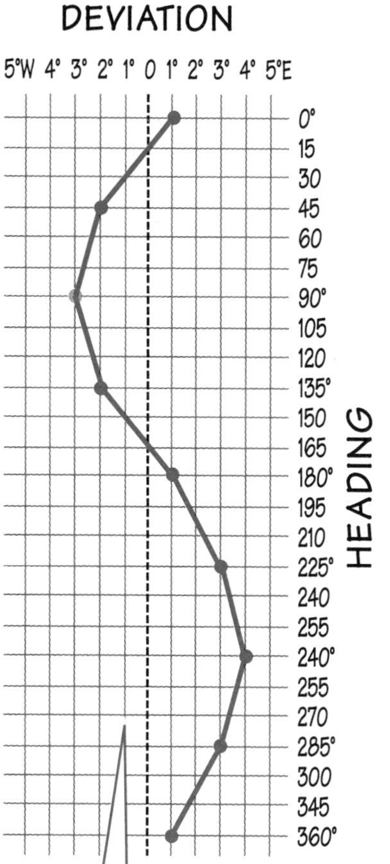

Plotting a Deviation Card
Using the 8 deviation figures gained from swinging ship (more if you wish, and/or others you have noted), plot a simple deviation card and join your readings together to form a graph. From this, you can readily read off deviation for intermediate headings.

If you discover deviation at any of the eight headings (000°, 045°, 090°, 135°, 180°, 225°, 270° and 315°), then work out whether deviation is "west" or "east" for a given heading using the "error west, compass best" mnemonic.

Example **Steering compass (C) 090°**
Bearing compass (M) 087°
Deviation (error) + 003° WEST
"Error west, STEERING compass best"

Example **Steering compass (C) 240°**
Bearing compass (M) 244°
Deviation (error) - 004° EAST
"Error East, STEERING compass least"

You may also choose to steer on 32 headings and plot each one on a table, rather than graphing the results of fewer headings. Either is equally effective. Making up the card may take a further 15 minutes, so to have a steering compass whose accuracy is fully understood has involved an investment of little over half an hour. Not much to spend on your primary navigation tool. Do this once a year, or when you take over an unknown yacht.

Distance Run — The Log

Distance run through the water is a vital element of position finding. If you cannot see anything except salt water, but you know how far you have come, you can always refer back to help establish your postion.

Distance is measured by an instrument called a *log*. The log may be electronic with a digital readout of speed as well as distance run, or a towed spinner activating a simple analog dial showing distance only. The accuracy of either depends on siting, maintenance and sea state. Do not expect perfection.

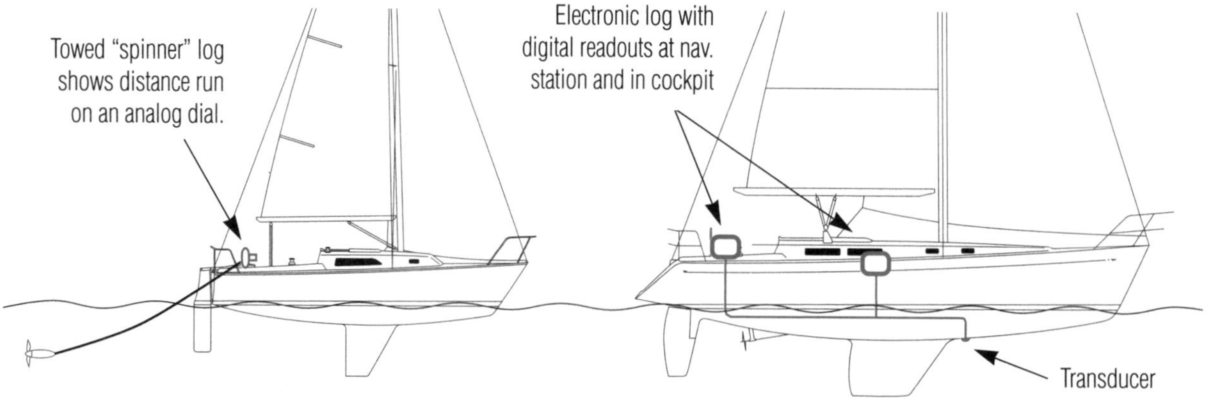

Just as on land or in the air, speed ties time to distance. Speed at sea is measured in *knots*, which are nautical miles per hour. If you are doing 4 knots, you are traveling four nautical miles through the water in one hour, and if you have covered three miles in half an hour, you are logging 6 knots.

Log Accuracy — It is important to know if your log reads accurately. If it doesn't, you need to determine whether its anomalies are temporary or permanent. For this purpose, authorities often set up *measured miles*, which are precisely defined geographic ranges that define exactly one mile over the ground. These measured miles are indicated on the chart, and are very useful in determining the accuracy of (*calibrating*) your log.

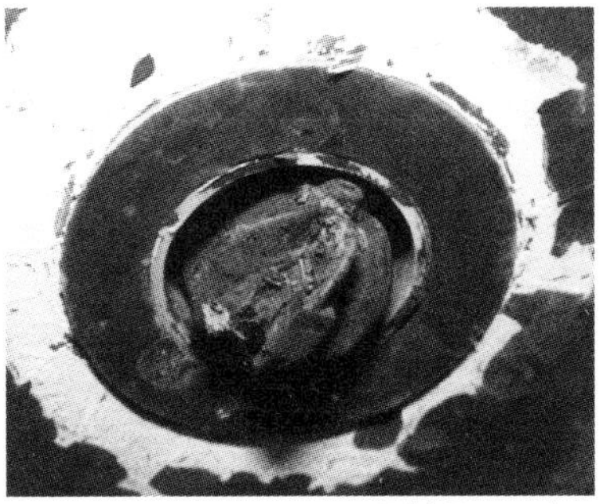

Ralph Naranjo photo

Logs of all kinds have a nasty habit of picking up weed and running "slow" or stopping. This is easy to rectify for a trailing log, but a through-hull impeller (shown above) may need to be withdrawn into the hull to be cleared. Some installations make provision for this. Others require steady nerves and a quick hand.

To calibrate your log, choose a calm day and motor steadily down a measured mile (or any other certain distance). Note your log reading at the start and then as you cross the second range. Now turn round and steer back over the mile doing the same again. Going both ways will compensate for any current that may be running. Add the two runs together, divide by two, and compare them with the charted distance you know you have really run. If you are in an area of strong tide, try to do this as near to slack water as possible. Running the distance twice and dividing by two will be flawed in a strong current. If there is some current, powering as fast as you can will minimize its effects. Once you know your standing log error, you can apply it each time and your corrected log readings should be accurate. On your own boat, you have the option of recalibrating the log.

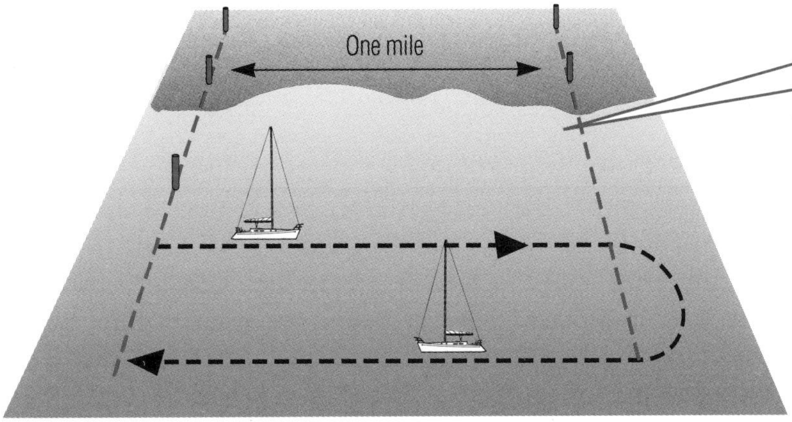

After motoring this measured mile in both directions and dividing the logged distance by two, the log of the boat (left) measured 2.05 miles instead of 2 miles. They now know the log over-reads by 0.025 per nautical mile, or 2.5 miles in a hundred.

Intuition — Finally, practice estimating your speed by looking over the side and watching your wake. If ever you lose your log, personal observation may be all you have left, so develop the skill *before* you need it!

Depth

After distance and direction, depth is the third navigational dimension. Inspect any chart. The water is dotted with figures called *soundings*, or known (charted) depths, measured from the sea floor to a "low tide" water level known as *Chart Datum*.

Until recently, Chart Datum on East Coast charts was taken as the line of *Mean Low Water*, or the average of all low tides compiled over a 19 year period. In areas experiencing two tides per day, one tide is normally lower than the other, in some places by a small amount, in others significantly. The lower of the two is designated *Lower Low Water*, and the 19-year average for Lower Low Water is called *Mean Lower Low Water*. For years Chart Datum on the West Coast has been based on Mean Lower Low water, making the charted depth slightly shallower. In recent years, the East coast has been also adapting Mean Lower Low Water as Chart Datum. If ever you are sailing "close to the mark" on depth and have used a tidal height calculation, be sure to check the chart near to the title to see what its Chart Datum actually is. Always remember that in areas affected by tide, there will usually be more water than the chart indicates except during extreme Low Water Spring tides (see Chapter 6, *Tidal Heights* for details on this).

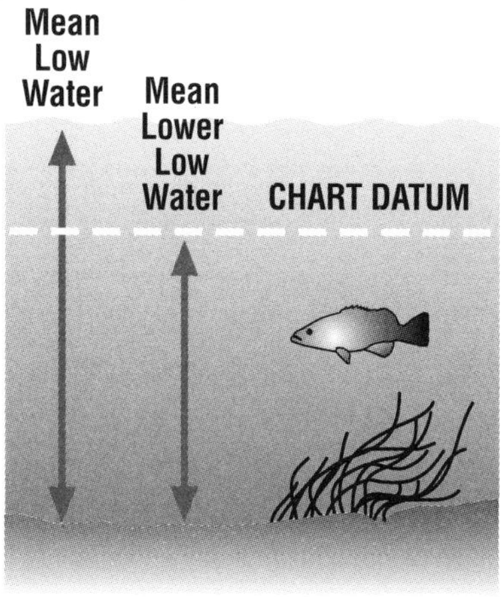

Traditionally, soundings are measured in *fathoms* (1 fathom = 6 feet) and feet. Shallower water is recorded as fathoms with a subscript indicating feet, for example, 4 fathoms and 3 feet is indicated as 4_3. Depths of less than a fathom, 3 feet, for example, are marked as: 0_3

Charts of comparatively shallow water may be sounded in feet only. If so, they will be clearly labeled as such.

There is a steady change in policy towards the internationally accepted meters as the unit for depth (and height). A metric chart is sounded in meters and tenths of a meter, using the same subscript system as fathoms do for feet. One meter is approximately equal to 3 feet, 3 inches. Every chart states the units of its soundings near the title. Read it, and take heed! It is all too easy to forget.

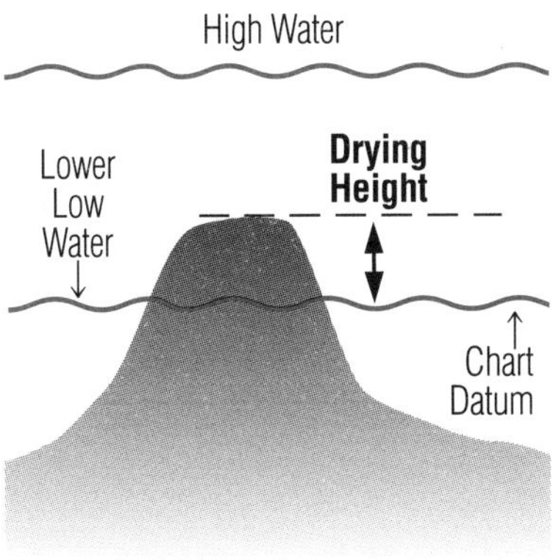

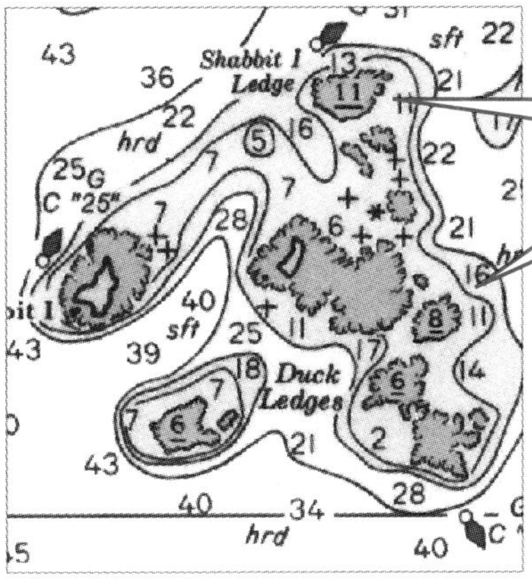

Drying heights are indicated by a sounding with a line underneath, describing the height of the area above Chart Datum at low water.

Drying heights — In tidal waters, some areas are above the surface at low water. These areas are charted as *drying*, and their soundings indicate height instead of depth. Heights are charted with a line underneath them, indicating *drying heights* above Chart Datum.

Measuring Depths — Depths are measured from the boat by an electronic device called a *fathometer,* or *depth sounder.* Usually, these read depth directly below their through-hull transducers, but some can be calibrated to read "depth below keel," or "overall depth" to the water surface (see below). Check to be sure that you know what yours is reading.

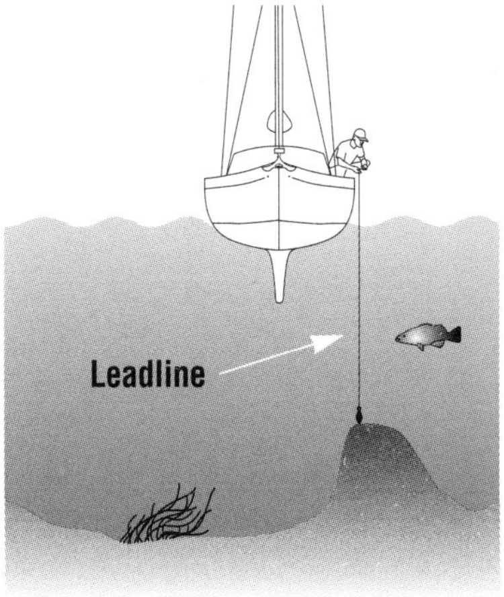

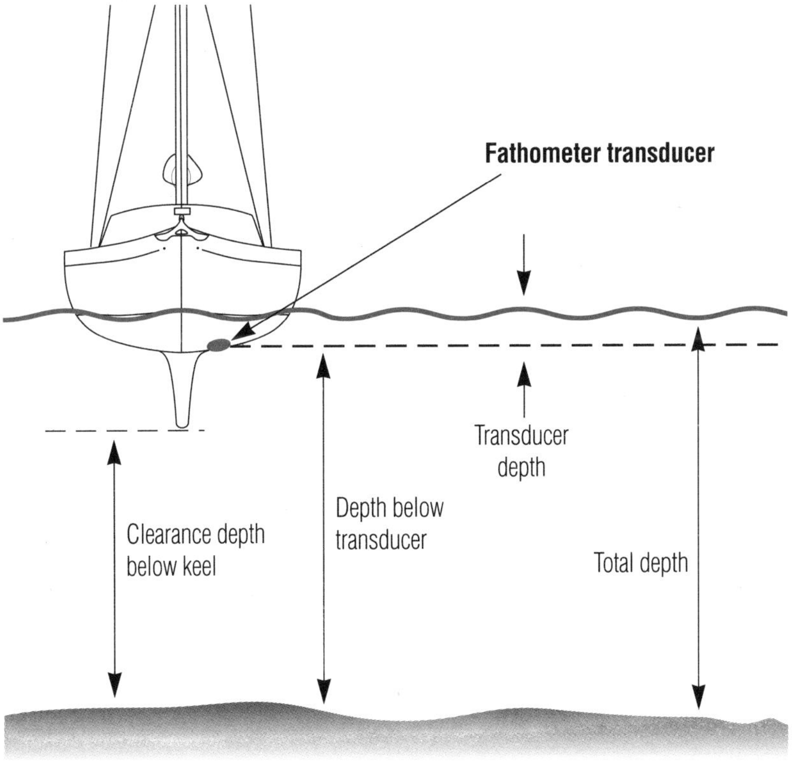

The *leadline* is an often under-appreciated device for determining depth. Difficult to use under way, it nonetheless provides an essential backup for failed electrics. Its real value is in sounding depths near the boat at rest with great accuracy. You can also use a leadline from a dinghy to sound a river entrance for which you have no chart, or to determine the true depth of water to check what the fathometer is reading. By this method you can ascertain the fathometer depth at which you will run aground — probably the most important depth of all to know. A leadline can be marked in meters or fathoms, but the vital mark which all lines should bear is the one which indicates exactly the depth at which the vessel will run aground.

CHAPTER 6 Tidal Heights

In most parts of the world, the rise and fall of tides is a crucial part of the inshore navigator's work. In the Bay of Fundy, Nova Scotia, for example, tides rise to heights of over 40 feet. Anyone who ignores such dramatic phenomena is going to be literally left high and dry. For many parts of the U.S., tide is far less dramatic, but a rise of 6 feet is not uncommon. If you sail in shallow waters, this is enough difference to render huge areas of apparently unusable water entirely navigable and vice versa.

Understanding the rise and fall of tides is a primary safety consideration. By using the tides one can often creep into a harbor which the chart suggests has an entrance too shallow to negotiate, or take a shortcut that will allow you to reach shelter before dark. Furthermore, if you are steering along a line of soundings in fog where the water is shallow and there is an appreciable tide rise, you must be able to compensate to carry out your fog strategy. In short, the competent pilot must be able to work tides. It's easier to do than you might imagine.

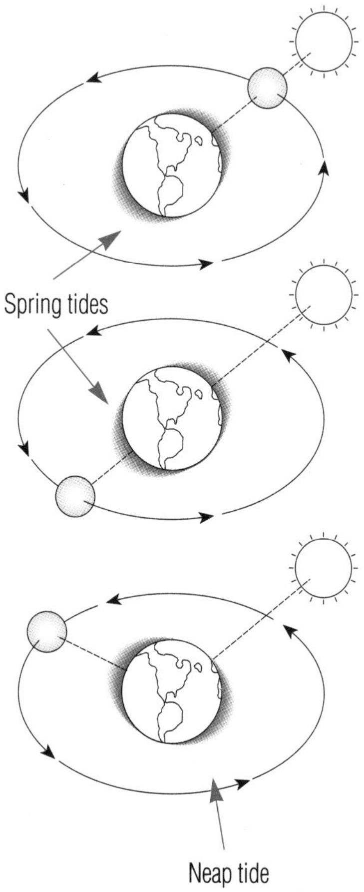

The first essential is to understand tidal terminology, so make sure you have absorbed the next few pages before reading further. Do not hesitate to refer back to the diagrams frequently. The most important diagram, once you understand the meaning of "Spring" and "Neap," (see left) is on page 34.

Why Tides Occur

Tides occur as a result of the combined gravitational pull of the Sun and the Moon. Even though it is much farther away, the Sun's effect is almost half that of the Moon because the Sun is massively heavier.

Put simply, the gravities of the Sun and the Moon act together when they are in a straight line with the Earth, regardless of whether or not they are both on the same side of the Earth. This condition occurs at full Moon and new Moon (dark moon), producing large tides known as *Springs*. The name "Spring" has nothing to do with the season. When Sun and Moon act against one another at half Moon, the tides are about half the size and are known as *Neaps*.

Most areas with significant tidal rise have two tides per day, i.e. two high waters and two low waters. Such tides are termed *semi-diurnal tides*. A *diurnal tide* occurs in zones with only one tide per day, often a tropical location of small rise and fall. Some places experience mixed tides (see charts at right).

A **diurnal tide** occurs in zones with only one tide per day, often a tropical location of small rise and fall. Most areas experience a **semi-diurnal tide** (two tides per day). Some (bottom) experience mixed tides.

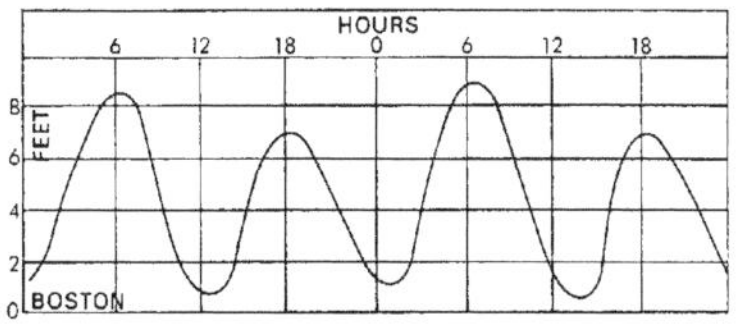

Semi-diurnal tide cycle – twice a day

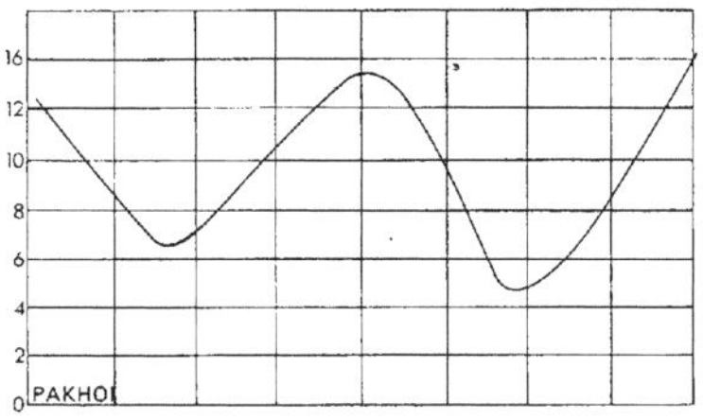

Diurnal tide cycle – one a day

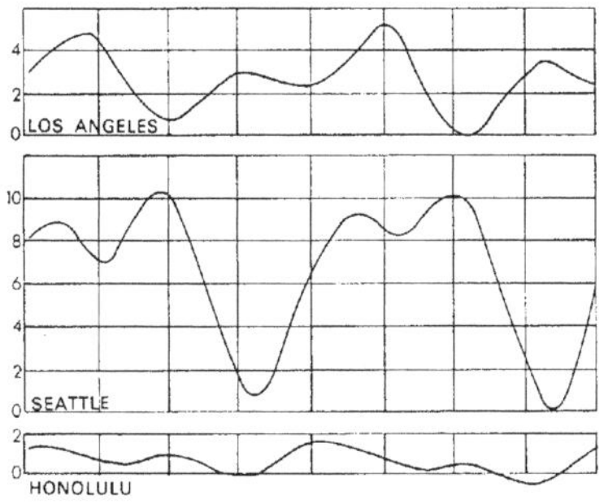

Mixed tide cycles

Tide Tables

The times and heights of high and low water for important commercial ports (Standard Ports) are given in *tide tables*. These are found in almanacs such as ELDRIDGE or REED'S, or in other locally produced publications. Most tide tables cover every day of the year. A typical tide table is shown below.

NEWPORT, RI

HIGH & LOW WATER 41°30'N 71°20'W

CORRECTED FOR DAYLIGHT SAVING TIME: APRIL 2 – OCTOBER 28

RHODE ISLAND

MAY

	Time	ft		Time	ft
1	0306	-0.1	16	0309	-0.8
	0936	3.5		0939	4.2
M	1456	0.0	Tu	1513	-0.7
	2153	4.0		2202	4.9
2	0335	0.0	17	0401	-0.7
	1017	3.3		1032	4.1
Tu	1530	0.1	W	1606	-0.5
	2233	3.8		2255	4.7
3	0407	0.1	18	0453	-0.6
	1100	3.2		1127	4.0
W	1605	0.3	Th	1702	-0.3
	2316	3.5		2351	4.4
4	0442	0.2	19	0549	-0.3
	1145	3.0		1225	3.8
Th	1645	0.4	F	1803	0.0
5	0000	[illegible]	20	0049	4.1
				[illegible]	-0.1

JUNE

	Time	ft		Time	ft
1	0345	0.2	16	0439	-0.5
	1035	3.3		1108	4.2
Th	1543	0.3	F	1651	-0.2
	2249	3.7		2330	4.5
2	0420	0.2	17	0530	-0.3
	1119	3.2		1203	4.1
F	1622	0.4	Sa	1749	0.1
	2332	3.5			
3	0457	0.3	18	0025	4.1
	1205	3.2		0622	-0.0
Sa	1706	0.5	Su	1259	4.0
				1853	0.3
4	0017	3.4	19	0121	3.8
	0538	0.3		0716	0.2
Su	1253	3.2	M	1356	3.9
	1755	0.6	☽	2004	0.6
5	0107	3.2	20	0218	3.5
	0625	0.3			
M	1345				

JULY

	Time	ft		Time	ft
1	0357	0.1	16	0502	-0.2
	1052	3.5		1136	4.3
Sa	1603	0.3	Su	1726	0.1
	2305	3.8		2357	4.1
2	0433	0.1	17	0546	0.0
	1135	3.5		1229	4.1
Su	1646	0.4	M	1820	0.4
	2348	3.6			
3	0512	0.2	18	0049	3.7
	1221	3.5		0631	0.3
M	1733	0.5	Tu	1322	3.9
				1918	0.7
4	0035	3.4	19	0142	3.4
	0557	0.2		0717	0.5
Tu	1311	3.6	W	1417	3.8
	1826	0.5	☽	2027	0.9
5	0127	3.3	20	0238	3.2
	0646	0.2		0808	0.7
			Th	1513	3.7
				[illegible]	1.0

AUGUST

	Time	ft		Time	ft
1	0447	-0.0	16	0014	3.6
	1151	3.9		0543	0.3
Tu	1714	0.3	W	1245	3.9
				1825	0.8
2	0008	3.6	17	0105	3.3
	0531	0.0		0623	0.6
W	1241	3.9	Th	1338	3.7
	1806	0.4	☽	1915	1.0
3	0100	3.4	18	0159	3.1
	0620	0.1		0708	0.8
Th	1337	3.9	F	1434	3.5
☾	1905	0.5		2019	1.2
4	0200	3.3	19	0257	2.9
	0716	0.2		0803	0.9
F	1438	4.0	Sa	1531	3.5
	2012	0.5		2149	1.2
5	0304	3.3	20	0355	2.9
	0821	0.2		0908	1.0
Sa	[illegible]				

Reed's Nautical Almanac

Before using a tide table, it is important to check the time standard used in the table. Often it is Zone Time, to which you may have to add an hour for Daylight Saving. Some local tables may be given in 12 hour clock time vs. 24 hour (shown).

The *height* of the tide is the amount by which the water stands above Chart Datum listed on a chart. If a tide table lists Low Water at a given date and time as 1.2 meters, and the charted depth for a specific location is 2.5 meters, the actual depth at Low Water will be:

2.5 meters	**(charted depth)**
+ 1.2 meters	**(tide)**
= 3.7 meters	**(depth at Low Water)**

If High Water is tabulated as 5.5 feet and the chart depth is 4 feet, the actual depth at High Water would be:

4.0 feet	**(charted depth)**
+ 5.5 feet	**(tide)**
= 9.5 feet	**(depth at High Water)**

Here is the same High Water figure applied to an area of the chart whose sounding shows as a drying height of 2 feet (2):

5.5 feet (tide)
– 2.0 feet (drying heights are negative depths)
= 3.5 feet depth of water at High Water

Sometimes, Low Water Springs depress sea level to a figure below that given on the chart. In these cases, the Low Water height is tabulated as a negative figure. If Low Water is – 0.8 meters and the charted depth is 2.8 meters, the depth will be:

2.8 meters
– 0.8 meters
= 2.0 meters

This illustration shows the principal tidal definitions. Note the differences between *neap* tide (left) and *spring* (right).

▶ Charted heights of land objects are given above Mean High Water, which is the average of all High Waters.

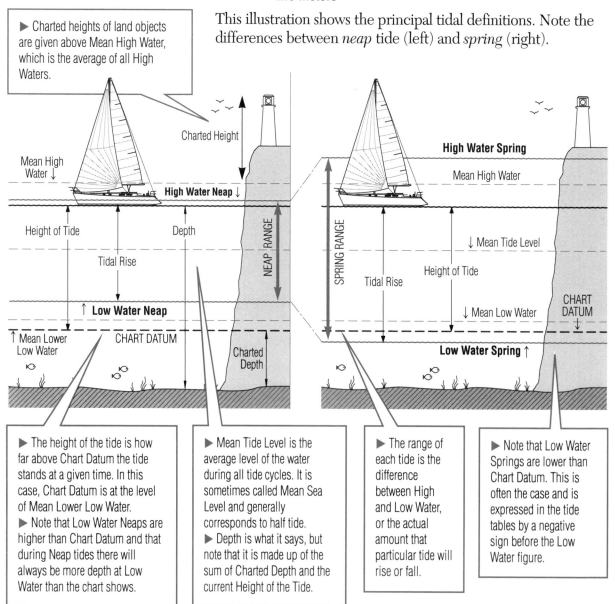

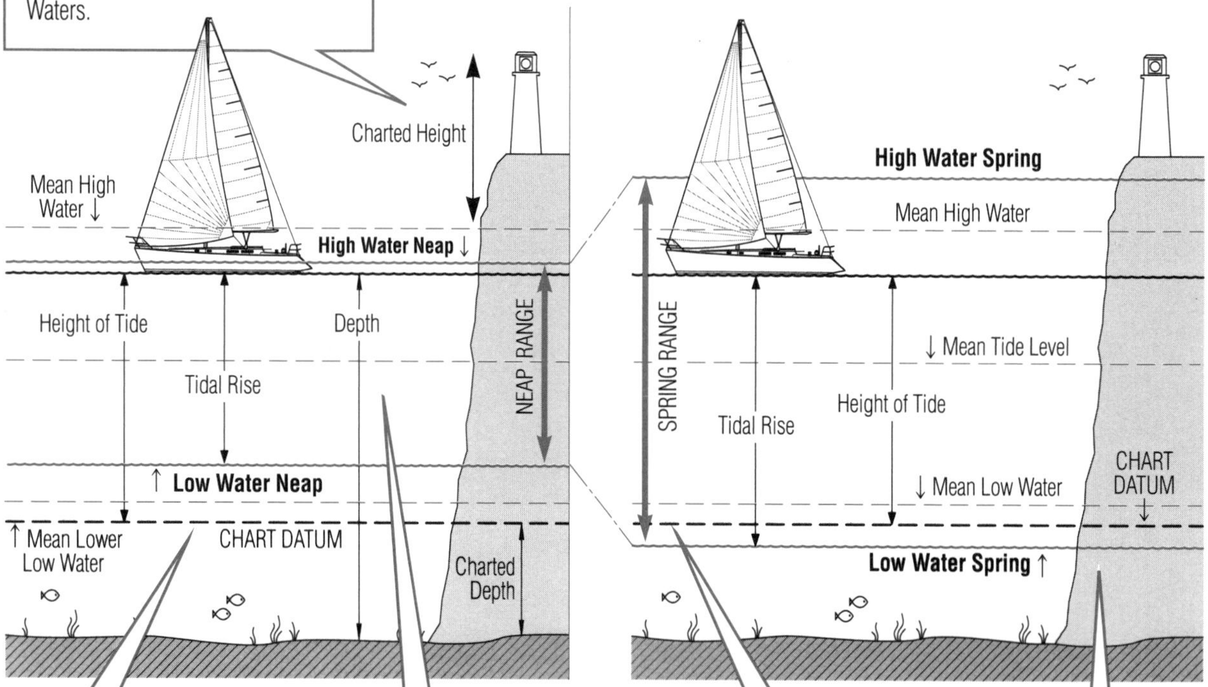

▶ The height of the tide is how far above Chart Datum the tide stands at a given time. In this case, Chart Datum is at the level of Mean Lower Low Water.
▶ Note that Low Water Neaps are higher than Chart Datum and that during Neap tides there will always be more depth at Low Water than the chart shows.

▶ Mean Tide Level is the average level of the water during all tide cycles. It is sometimes called Mean Sea Level and generally corresponds to half tide.
▶ Depth is what it says, but note that it is made up of the sum of Charted Depth and the current Height of the Tide.

▶ The range of each tide is the difference between High and Low Water, or the actual amount that particular tide will rise or fall.

▶ Note that Low Water Springs are lower than Chart Datum. This is often the case and is expressed in the tide tables by a negative sign before the Low Water figure.

Secondary Stations

Stations of secondary commercial importance are not fully tabulated in tide tables because to do so would require an almanac thousands of pages thick. Instead, such places are dealt with as sub-stations of the most convenient standard port. Secondary stations are found in the almanac (or the local tables) next to their standard port, together with information known as *tidal differences*. Height information given for that day at the standard port may be in the form of a ratio to find the equivalent data for the secondary station, or as a height difference rather than a ratio (see Appendix). If in doubt, take the time to read the directions in the publication carefully. They are generally crystal clear. Time differences are self-explanatory.

	Time	Height
HW Portland	0214 hours	9.40 feet
Differences - Steele Harbor Island	- 28 minutes	x 1.27 ratio
HW Steele Harbor Island	= 0146 hours	= 11.90 feet

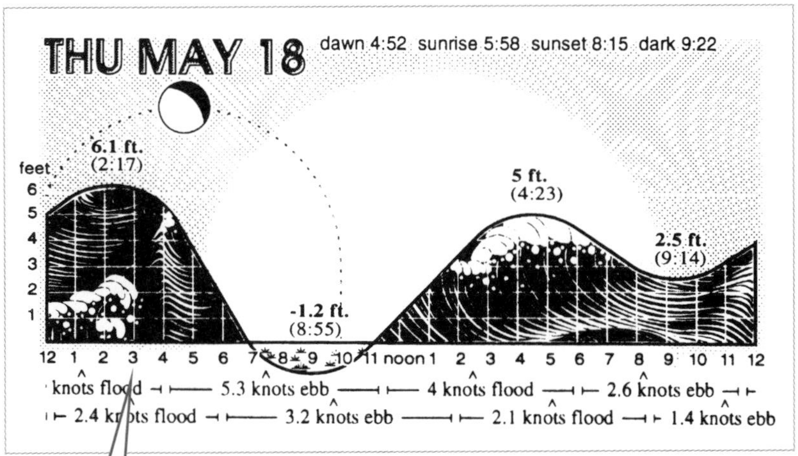

Tidal Heights between High and Low Water

To work out the height of tide at a time between High and Low Water, you have two choices. In some tide tables you may find a graphical representation of the rise and fall on each day (see above). If this is the case, reading the graph is easy and will usually give you the height of the tide above Chart Datum. Thus: at 0300 the tide will have risen to a height of 6 feet. Notice that the next Low Water is below Chart Datum as the spring tide advances.

If there is no convenient graph or tide table, an alternative is to use the *Rule of Twelfths*. This is based on an even rise and fall of tide and expresses a smooth progression in numerical terms. The rule figures how much of the total tidal range will have risen or fallen at a given time. The answer it gives must therefore be added to the height of Low Water. A glance at the diagram on page 34 shows that Tidal Height = Rise + Low Water Height.

The Rule of Twelfths

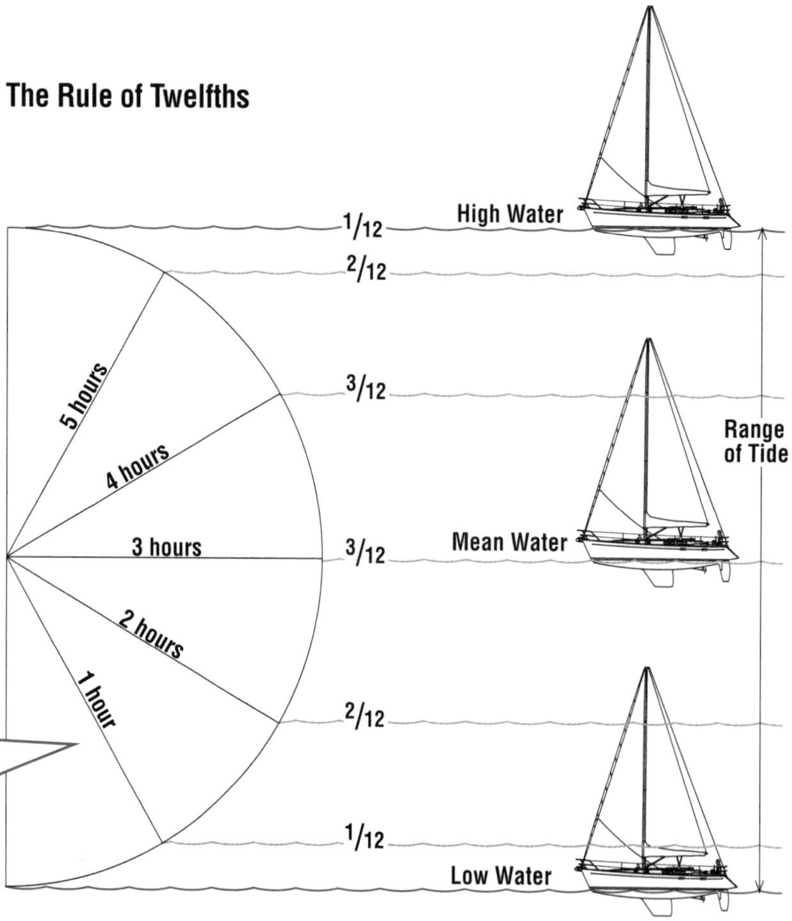

Example:
Low Water is at 0300, height 2 ft.
High Water is at 0900, height 8 ft.
Find the tidal height at 0700

- Tide range is 8 ft - 2 ft. = 6 ft.
- 1/12 of range is 6 ft ÷ 12 = 6 in.
- At 0700, 9/12 of the range will have risen (1/12 + 2/12 + 3/12 + 3/12 = 9/12). 6 ft x 9/12 = 4 ft. 6 in. This is the rise of tide.
- Add this figure to the height of Low Water to find the Height of Tide, i.e., 4 ft 6 in + 2 ft = 6 ft. 6 in.

To use the rule, first calculate the range of that particular tide. This is simply the difference between High and Low Water. (If you are figuring for a secondary port, apply the rule to the High and Low Water figures you have worked out for it, rather than using those for the standard port.) Now divide the range by 12 and apply the Rule of Twelfths as shown in the illustration. This gives the rise of tide at the hour you want. 1/12 of the range rises in the first hour, 2 in the second, 3 in the third, 3 in the fourth, 2 in the fifth and 1 in the sixth hour of the tide. The water falls at a corresponding rate on the ebb. Remember the mnemonic "1-2-3-3-2-1." Add this figure to the Low Water height, and you have the height of tide for the time of your calculation. See example at left.

Practical Tidal Height Calculations — Looking at a chart and asking yourself how much water there will be at a place at a given time is a logical matter of adding the tidal height to the charted sounding. Often, however, you may need to work the sum backwards. Suppose, for example, you intend to enter a harbor with a shallow entrance for which you require an additional 4 ft 6 in of water for a safe arrival. The question then becomes: *At what time will the tide have risen to a height of 4 ft. 6 in.?*

For many tidal questions it is helpful to draw a diagram of what you require. Once you know what you are looking for, the answer to the sums is found by entering the graph or the rule of twelfths with a height and extracting a time.

Example:

- ▶ Charted depth is 3 feet
- ▶ Safe depth for entering is decided to be 7ft. 6in.
- ▶ Therefore 4ft. 6in. is the least possible height of tide to enter harbor with a sensible clearance (7ft. 6in. – 3ft. = 4ft. 6in.)
- ▶ Low Water...3ft. at 0900
- ▶ High Water...7ft. at 1500

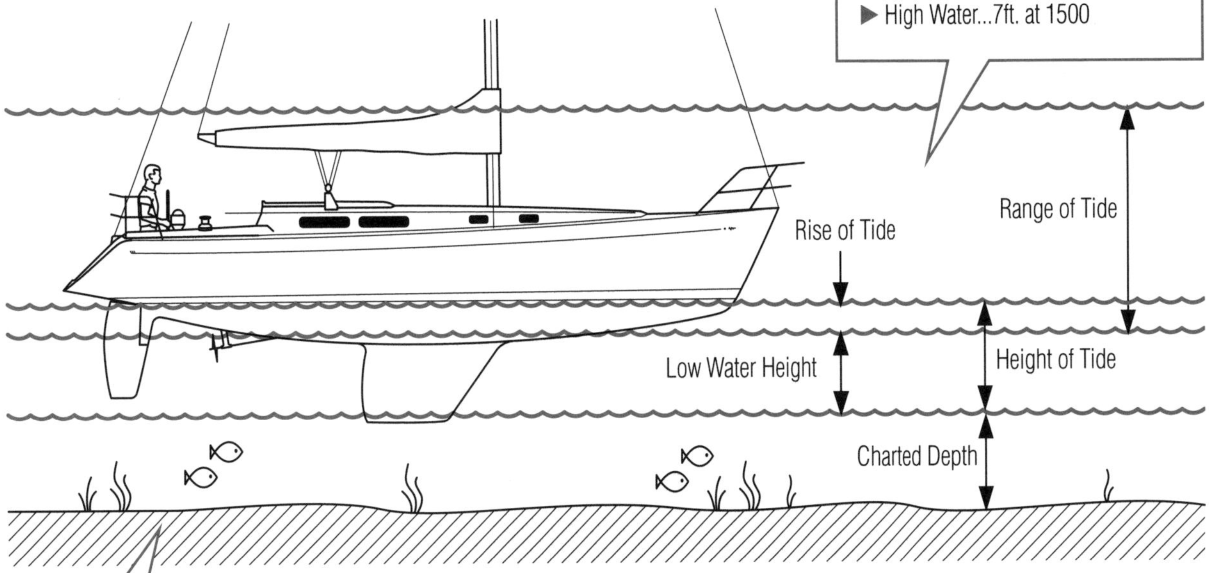

What is the earliest time at which I can safely come in?

- ▶ Range = 4feet (7ft. – 3ft.).
- ▶ 1/12 of range = 1/3 foot, or 4 inches.
- ▶ Rise of Tide we are looking for is the Height of Tide minus the Low Water Height, i.e., 4ft. 6in. – 3ft. = 1ft. 6in.
- ▶ 4 inches (1/12) divides into 1ft. 6in. (18in) just under 5 times, therefore we are looking for a rise of 5/12
- ▶ 1/12 rises in hour 1+ 2/12 rises in hour 2...so by 1100 we have 3/12 "up."
- ▶ We add 2/12 to allow a small safety margin, but 3/12 will rise in hour 3.
- ▶ So during hour 3, our last 2/12 will come up in approximately 2/3 of an hour (40 minutes).
- ▶ The required rise will therefore have arrived by 1140.

Anchoring on a Falling Tide — In this case, the yacht has anchored and the skipper wonders whether it will still be afloat at Low Water.

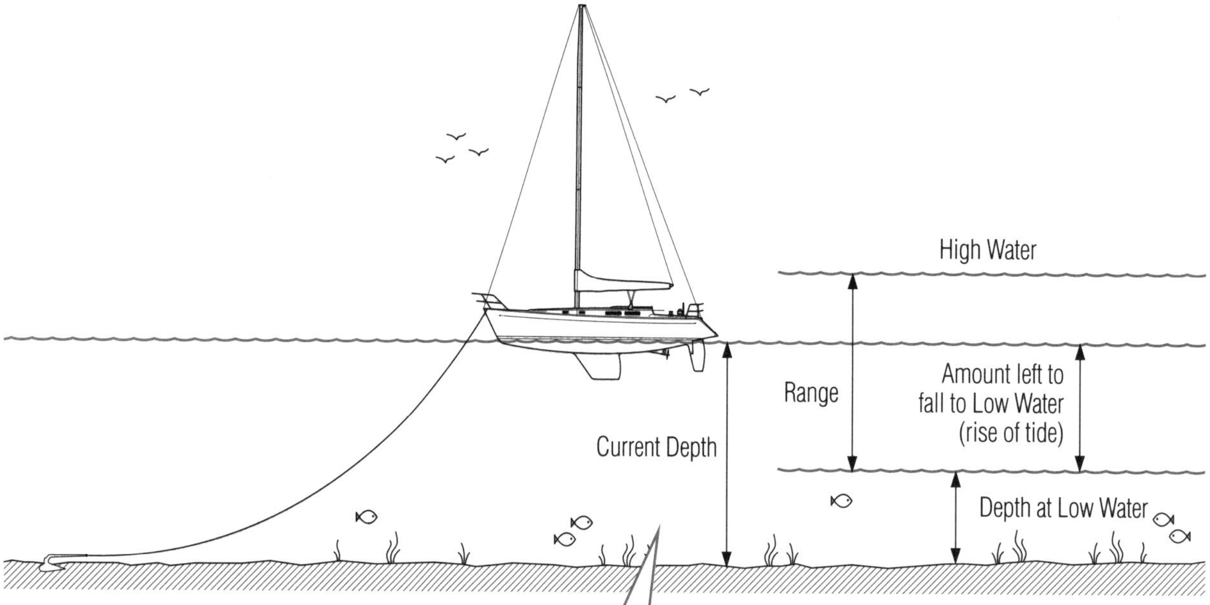

In this diagram it is clear that the current depth on the fathometer represents the Rise of Tide + the Depth at Low Water. The skipper needs to know what the depth will be at Low Water. All that is necessary, therefore, is to work out how much tide there is left to fall. This can be done either by the Rule of Twelfths or by reference to a tidal height graph if there is one. The figure is subtracted from the current depth and the result is the Low Water depth in question.

NOTE: *Charted depth does not come into this equation.* This is because you are unlikely to find an accurate sounding exactly where you lay your anchor. The only depth you are certain of is the one from your own fathometer. It makes sense to work this system in advance to decide what depth to anchor in when approaching a tidal anchorage.

Safe Clearances

In any coastal sailing there will be times when you operate in shoal water as a matter of choice. The amount of clearance you leave under your keel must depend upon a prudent assessment of such conditions as sea state, whether you are on a lee shore (a shore with the wind blowing onto it) or a safer weather shore and, to be absolutely cynical, the nature of the bottom. For example, it may be acceptable to dice with the mud on a weather shore on a rising tide. It can never be acceptable to be too close to a rock bottom on a falling tide on a lee shore.

It is imperative to be aware that all tidal information comes in the form of *predictions* which are no more than they say they are. Always allow a margin for error. Make your calculations as carefully as you can, then, starting from that position of strength, assess the likely accuracy of the prediction on the day. Prolonged onshore winds will often push a tide up. High atmospheric pressure will depress it at the rate of 1 foot per 35 millibars or so, starting from a standard pressure of 1013 millibars. Similarly, low pressure will raise the tide, as can large amounts of river water entering bays after heavy rainfall.

The list of extraneous effects is long, but it should not mean you are to walk away from the tide and say "that's not for me." Tide is a fact of life with which all serious sailors must work. An ability to do so brings great satisfaction and places the good navigator apart from the crowd.

CHAPTER 7 Tidal Currents

Currents are horizontal movements of the water. Ocean currents are generally predictable, running in one direction for long periods. Coastal currents that result from the gravitational pull of the Moon and Sun usually change their direction twice each day. Currents are also affected by water depth, surrounding geography and time of year.

Effects of Water Movement

Generally, currents run more strongly in deep water than in shoal (shallow), so if you are sailing up a river with the current you are better off in the middle. When you wish to work against the stream, your preferred course is to hold close to the shore. The same thing applies on the coast. There is less current "along the beach" than will be found further out in deep water.

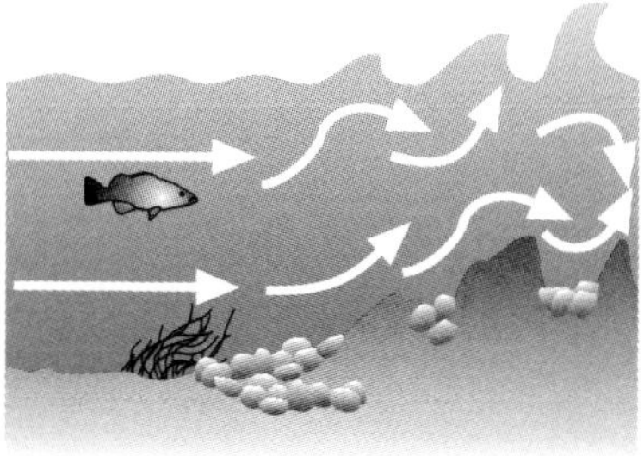

Current running into shallow water with a rough seabed will produce waves of increased height and steepness.

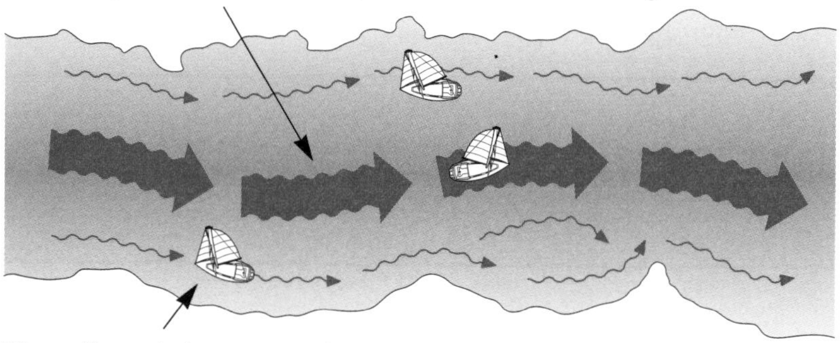
When sailing with current, head for deeper water where current is stronger
When sailing against current, stay close to shore where current is weaker

Where current runs in shallow water over a rough seabed, the sea surface will be broken, producing waves of greater height and steepness than would otherwise be expected.

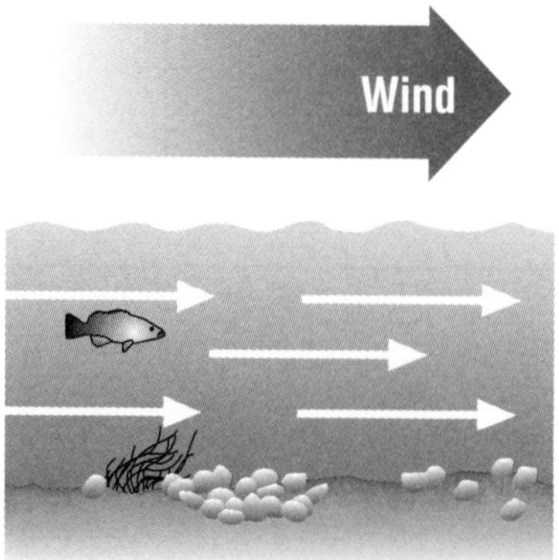
Wind

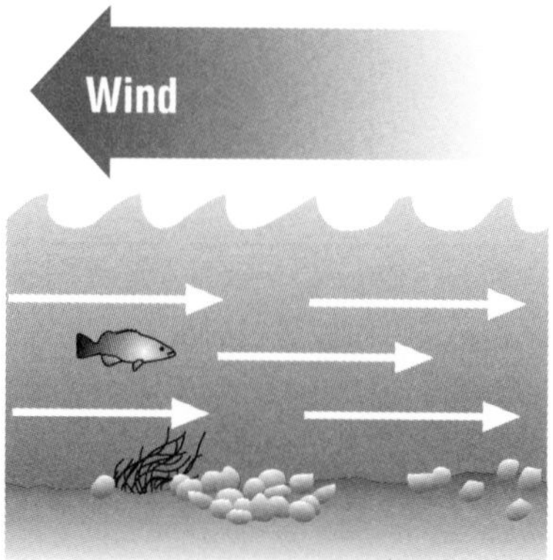
Wind

Expect rough seas where a current runs to windward. The Gulf Stream is notorious for this, but the same thing happens with tidal currents. Ask anyone who has beaten out through the Golden Gate on the ebb in a hard west wind. It is most important to keep this in mind when sailing in current affected waters.

Non-tidal Currents (Streams)
These generally run along a coast, with the Gulf Stream as the outstanding example. Their position may vary considerably, sometimes with the seasons, sometimes at random.

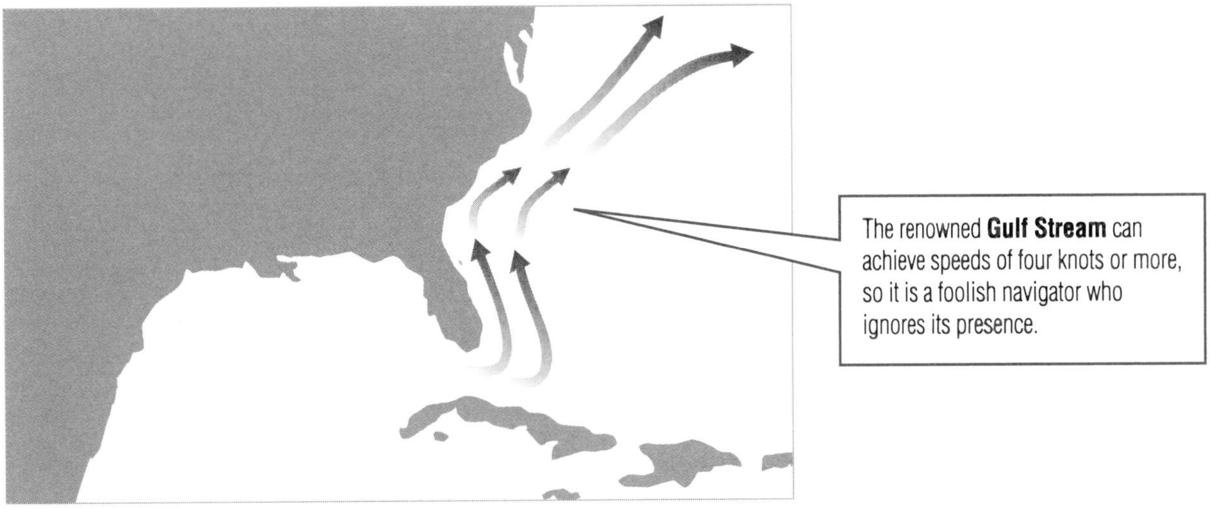

The renowned **Gulf Stream** can achieve speeds of four knots or more, so it is a foolish navigator who ignores its presence.

Major currents will be noted on the relevant charts, but for fuller information you should consult the Coast Pilot, an ocean pilot chart, or your pilot book. If you are going foreign, you will have to look it up in the relevant volume of the Sailing Directions. Up-to-the-minute information on current movement may be available in some areas from coast radio stations or the Coast Guard.

Tidal Currents
Significant tides generally ebb and flow twice each day. When the resulting currents are noticeable, navigators ignore them to their inconvenience and sometimes peril. Rates of tide vary with the phase of the Moon. Full Moon and new Moon generate full-rate currents known as Spring Tide. Half-moon conditions are much gentler and are called Neaps. If you want to know why this should be, read *Chapter 6, Tidal Heights.*

Information on tidal currents, or streams, can be found in various places. Nautical almanacs such as REED'S or ELDRIDGE offer data for a whole coast, but if most of your sailing is in a specific locality where tides run strongly, there may well be a local publication which is easy to read and filled with useful detail. Data may take the form of a current table or it may be found in a series of tidal current charts with arrows to indicate the strength and direction of the water movement.

Current Tables are easy to read, giving the general direction and specific strength in knots for each maximum current at the standard tidal station. Times of slack water and of maximum current are tabulated. For intermediate times, the speed of the current can be calculated using a table titled, *Speed of Current at Any Time*, in Reed's Nautical Almanac or International Marine's Tidal Current Tables (see below or Appendix).

Sometimes, current tables refer to currents at a **secondary station.** The rules for these are the same as for "secondary ports" in tidal height calculations (*see Chapter 6*).

THE RACE, LONG ISLAND SOUND

CURRENT TABLE 1996 — **FLOOD 302° EBB 112°**

CORRECTED FOR DAYLIGHT SAVING TIME: APRIL 7 – OCTOBER 26

JULY

Day	Slack time	Max time	Fld knots	Ebb knots
1 M		0255		4.2
	0613	0903	3.6	
	1220	1518		3.7
	1823	2120	3.8	
2 Tu	0035	0345		4.3
	0702	0953	3.7	
	1311	1610		3.8
	1916	2211	3.8	
3 W	0126	0436		4.3
	0752	1044	3.7	
	1403	1702		3.8
	2011	2304	3.6	
4 Th	0219	0527		4.1
	0843	1136	3.7	
	1456	1756		3.7
	2107	2358	3.4	
5 F	0314	0620		3.9
	0936	1229	3.5	
	1552	1851		3.6
	2207			
6 Sa		0055	3.1	
	0412	0715		3.5
	1031	1324	3.2	
	1649	1949		3.4
	2309			
7 Su ☽		0154	2.8	
	0512	0813		3.2
	1128	1422	3.0	
	1749	2049		3.2
8 M	0013	0257	2.5	
	0616	0912		2.9
	1228	1523	2.8	
	1849	2149		3.1
	17	0403	2.4	
	19	1013		2.7
	28	1626	2.6	
	48	2250		3.0
	18	0510	2.3	
	21	1112		2.6
	26	1727	2.6	
	44	2347		3.0
	15	0611	2.3	
	18	1208		2.6
	21	1822	2.6	
	36			
		0039		3.1
	06	0704	2.4	
	10	1259		2.6
	11	1910	2.6	
	23			
		0126		3.1
	53	0748	2.4	
	56	1345		2.6
	56	1952	2.6	
	07			
		0209		3.2
	35	0826	2.5	
	39	1428		2.7
	38	2030	2.6	
	47			
		0249		3.2
	14	0900	2.6	
	19	1508		2.7
	18	2107	2 6	
16 Tu	0025	0328		3.2
	0650	0935	2.6	
	1256	1547		2.8
	1856	2144	2.6	
17 W	0101	0406		3.2
	0726	1010	2.6	
	1332	1626		2.8
	1934	2222	2.6	
18 Th	0137	0445		3.1
	0800	1047	2.7	
	1408	1706		2.8
	2012	2301	2.6	
19 F	0213	0524		3.0
	0836	1126	2.6	
	1445	1747		2.7
	2053	2343	2.5	
20 Sa	0251	0605		2.8
	0913	1207	2.6	
	1523	1831		2.7
	2137			
21 Su		0027	2.3	
	0333	0649		2.7
	0953	1251	2.5	
	1605	1917		2.6
	2225			
22 M		0115	2.2	
	0420	0736		2.5
	1038	1339	2.5	
	1652	2008		2.6
	2318			
23 Tu ☾		0207	2.2	
	0513	0828		2.4
	1128	1431	2.5	
	1745	2102		2.7
24 W	0017	0303	2.1	
	0613	0924		2.4
	1224	1526	2.5	
	1843	2200		2.8
25 Th	0118	0403	2.2	
	0717	1023		2.5
	1323	1625	2.6	
	1944	2258		3.0
26 F	0219	0503	2.4	
	0821	1122		2.7
	1424	1724	2.9	
	2043	2356		3.4
27 Sa	0317	0602	2.7	
	0921	1220		3.0
	1523	1821	3.1	
	2140			
28 Su		0051		3.7
	0412	0659	3.1	
	1017	1315		3.3
	1620	1917	3.4	
	2235			
29 M		0144		4.0
	0503	0752	3.4	
	1110	1408		3.6
	1715	2011	3.7	
	2328			
30 Tu ○		0236		4.2
	0553	0844	3.7	
	1202	1500		3.9
	1808	2103	3.8	
31 W	0020	0326		4.4
	0642	0934	3.8	
	1252	1552		4.1
	1901	2155	3.9	

AUGUST

Day	Slack time	Max time	Fld knots	Ebb knots
1 Th	0111	0416		4.3
	0731	1024	3.9	
	1343	1643		4.1
	1954	2246	3.7	
2 F	0202	0506		4.2
	0820	1114	3.8	
	1434	1734		3.9
	2048	2338	3.5	
3 Sa	0255	0557		3.9
	0911	1204	3.5	
	1526	1827		3.7
	2144			
4 Su		0031	3.2	
	0349	0650		3.5
	1004	1257	3.2	
	1621	1922		3.4
	2243			
5 M		0127	2.8	
	0447	0745		3.1
	1059	1352	2.9	
	1718	2019		3.1
	2344			
6 Tu ☽		0227	2.4	
	0548	0843		2.7
	1158	1451	2.6	
	1818	2119		2.9
7 W	0047	0331	2.2	
	0651	0943		2.5
	1259	1553	2.4	
	1918	2219		2.8
8 Th	0150	0440	2.1	
	0753	1044		2.4
	1359	1657	2.3	
	2016	2318		2.7
9 F	0248	0545	2.1	
	0852	1142		2.4
	1456	1756	2.3	
	2110			
10 Sa		0013		2.8
	0340	0639	2.2	
	0944	1234		2.4
	1547	1847	2.4	
	2159			
11 Su		0101		2.9
	0427	0723	2.3	
	1031	1321		2.6
	1634	1929	2.5	
	2243			
12 M		0144		3.0
	0508	0800	2.5	
	1113	1403		2.7
	1716	2007	2.6	
	2323			
13 Tu		0224		3.1
	0546	0834	2.6	
	1151	1443		2.8
	1755	2043	2.7	
14 W ●	0001	0302		3.2
	0622	0907	2.7	
	1227	1521		2.9
	1832	2119	2.7	
15 Th	0037	0339		3.2
	0656	0942	2.8	
	1302	1559		3.0
	1909	2156	2.7	
16 F	0112	0417		3.2
	0729	1017	2.8	
	1335	1637		3.0
	1945	2234	2.7	
17 Sa	0147	0455		3.1
	0802	1055	2.8	
	1410	1717		3.0
	2023	2314	2.7	
18 Su	0223	0534		2.9
	0838	1135	2.8	
	1446	1758		2.9
	2105	2357	2.6	
19 M	0303	0617		2.8
	0917	1218	2.7	
	1527	1844		2.9
	2151			
20 Tu		0044	2.4	
	0348	0704		2.6
	1002	1305	2.6	
	1614	1934		2.8
	2244			
21 W ☾		0136	2.3	
	0441	0756		2.5
	1054	1358	2.5	
	1709	2030		2.8
	2344			
22 Th		0233	2.2	
	0543	0854		2.4
	1153	1457	2.5	
	1812	2130		2.8
23 F	0049	0335	2.3	
	0651	0957		2.5
	1258	1559	2.6	
	1918	2232		3.0
24 Sa	0153	0438	2.5	
	0758	1059		2.7
	1405	1702	2.8	
	2023	2333		3.3
25 Su	0254	0541	2.8	
	0901	1200		3.1
	1508	1803	3.1	
	2123			
26 M		0031		3.6
	0351	0639	3.1	
	0958	1257		3.5
	1606	1901	3.4	
	2220			
27 Tu		0125		4.0
	0443	0734	3.5	
	1051	1351		3.8
	1701	1956	3.7	
	2313			
28 W ○		0217		4.2
	0533	0825	3.8	
	1142	1442		4.1
	1754	2047	3.8	
29 Th	0004	0307		4.3
	0621	0914	3.9	
	1231	1532		4.2
	1845	2137	3.9	
30 F	0054	0355		4.2
	0708	1002	3.9	
	1319	1621		4.2
	1935	2226	3.7	
Sa	0143	0443		4.0
	0756	1049	3.7	
	1408	1710		4.0
	2026	2315	3.5	

TIME M… AN 75°W — 00… s midnight, 1200h is noon.

199… T COAST 3.69

Tides & Currents

Reed's Nautical Almanac

TABLE 3.—SPEED OF CU…

TABLE A…

…tween slack and desired time (h. m.)	Interval between s… h. m. 1 20	h. m. 1 40	h. m. 2 00	h. m. 2 20	h. m. 2 40	h. m. 3 00	h. 3…
	ft.	ft.	ft.	ft.	ft.	ft.	ft.
0 20	0.4	0.3	0.3	0.2	0.2	0.2	0…
0 40	0.7	0.6	0.5	0.4	0.4	0.3	0…
1 00	0.9	0.8	0.7	0.6	0.6	0.5	0…
1 20	1.0	1.0	0.9	0.8	0.7	0.6	0…
1 40	----	1.0	1.0	0.9	0.8	0.8	0…
2 00	----	----	1.0	1.0	0.9	0.9	0…
2 20	----	----	----	1.0	1.0	0.9	0…
2 40	----	----	----	----	1.0	1.0	1…
3 00	----	----	----	----	----	1.0	
3 20	----	----	----	----	----	----	

As with all tidal information, when a current chart or table gives a maximum current rate, you can take it that while things change continuously, if you assume that the stated rate begins at half an hour before maximum current and ends half an hour after it, you will not go far wrong. It is a mistake to believe that the stated rate begins at the stated time and continues until 1 hours later. (The current rate may refer to slack water instead of maximum current.)

Example:

To determine the velocity and direction of the current at The Race, Long Island Sound for 1330 hours, August 29, find the following information on the tidal current table (see above right).

Time of slack water = 1231 hours

Time of maximum current = 1532 hours

Velocity and direction of maximum current = 4.2 knots and ebbing

To enter the *Speed of Current at Any Time* table, we need to have the interval between slack and maximum current:

1532 - 1231 = 3 hours, 1 minute

and the interval between slack water and the desired time:

1330 - 1231 = 59 minutes

With these values, we get the factor 0.5 from the *Speed of Current at Any Time* table (see above left). The maximum current is multiplied by this factor to get the current at 1330 hours.

4.2 knots x 0.5 = 2.1 knots and ebbing

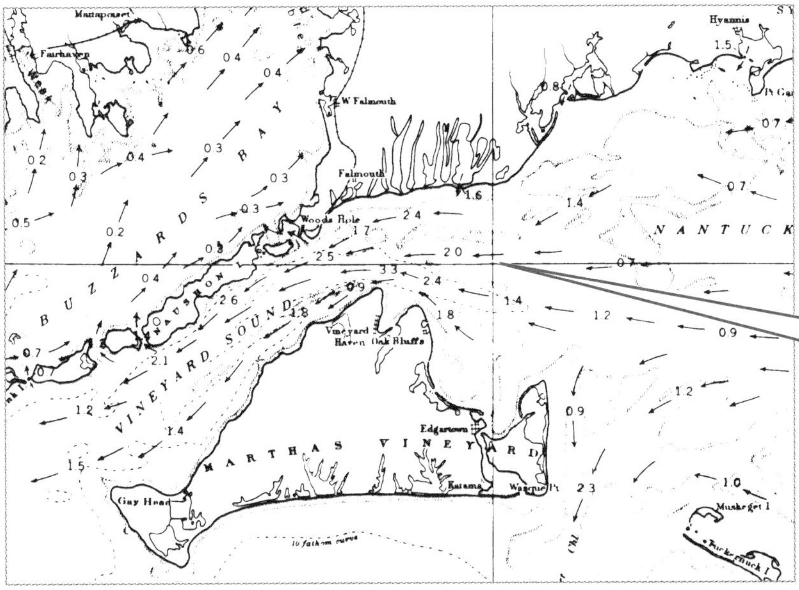

Reed's Nautical Almanac

Example:
Current for this hour is given on the tidal chart as 2 knots.
Spring Range is 4 feet.
Today's range is 3 feet.
Today's current at this hour is 2 divided by 4 = 1/2 x 3 = 1 1/2 knots.

Tidal Current Charts come in sets of 11 or 12, related to "hours before or after maximum current or slack water at the local standard port. Rates are given for spring (big) tides. For lesser tidal ranges, divide the rate by the spring range, then multiply it by the range for the day. (See *Chapter 6 - Tidal Heights* - for a full explanation of the term "range," which is the difference between high and low water for a particular tide.) In practice, you can often make a satisfactory guess. Enter the tide tables for the standard port to read off the time of high water, note its range and away you go.

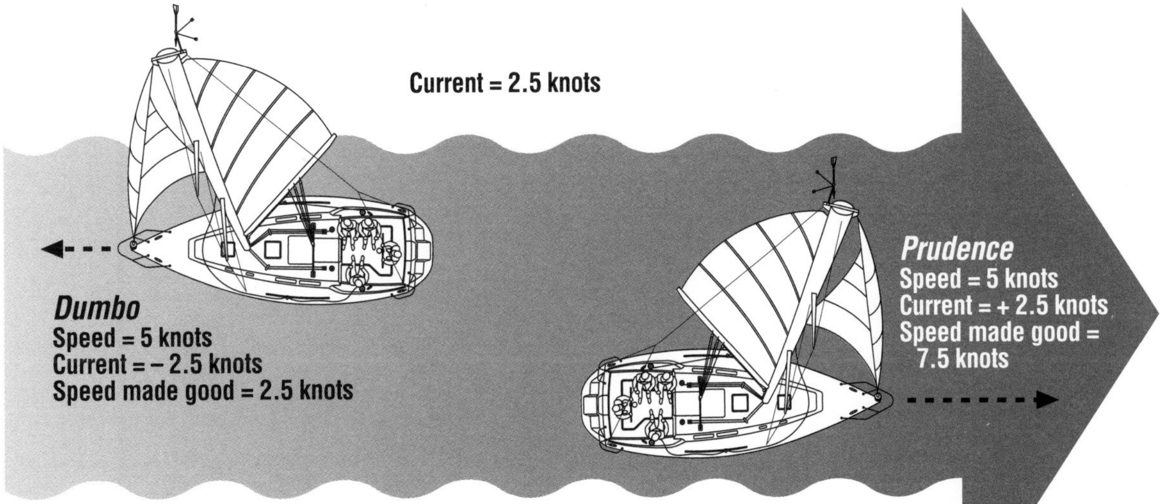

Always try to sail with the current if you possibly can. This illustration shows two boats both reaching. Taking into account the boat speed of a 34-foot cruiser in a moderate breeze as 5-6 knots, a 2.5 knot current (see above) can have a dramatic effect on progress. *Prudence* is sailing at 5 knots with a favorable 2.5 knot current for a total of 7.5 knots over the bottom. *Dumbo* is sailing at the same speed against the current for a total of 2.5 knots over the bottom. Both speed logs will read 5 knots, but *Dumbo* may as well have stayed home!

Personal Observation

No tide table is as accurate at predicting the turn of current as your own observation. Look at buoys and lobster pots, or anything else which is attached to the seabed. The water streaming past may be doing what it is not supposed to be doing, but nothing the tide tables can ever say will change that. Look at other boats' progress inshore of you to spot early eddies on the turn of the tide, and note the way anchored ships are lying. Observe creatively, and read the signs.

Anne Martin photo

Your most reliable and accurate indicator of current is your environment. A buoy is a useful indicator of the direction and strength of current (see left). A piling that is dry above the water level indicates that the tide is coming in (*flooding*). When the piling above the water is wet, the tide is going out (*ebbing*). A dry beach is indicative of a rising tide; a wet beach is a sign of a falling tide.

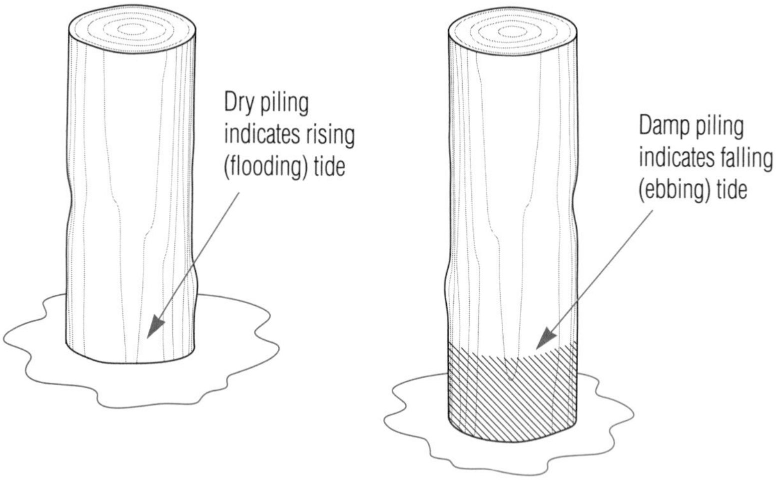
Dry piling indicates rising (flooding) tide
Damp piling indicates falling (ebbing) tide

NAV. NOTES...

Whenever you leave a dock, mooring or anchorage, check that the tidal current is doing what you expect. Entire passages have been planned using maximum ebb instead of maximum flood with interesting results! Good navigators always check their input, no matter how basic it is!

Local Knowledge

In many areas, weather conditions both nearby and far away can affect the accuracy of the official tidal current predictions. For example, a great deal of rain inland will often prolong the ebb in a bay or estuary. It might also pep up the rate.

Well known anomalies will be noted in the Coast Pilots, but for more intricate detail, ask a local boatman, preferably one who is too busy working the eddies to have more than a few moments to stop and shoot the breeze.

CHAPTER 8 Planning a Course to Steer

In Chapter 1 it was noted that navigation consisted of knowing where you are and being able to work out how to sail from there to where you want to go. In many North American locations, currents are slack, making the latter demand easily satisfied. Even when crosscurrents are to be considered, the job remains logical and can be dealt with in a simple visual way.

Basic Course to Steer

You will recall that in Chapter 2, the technique of using a chart plotter or parallel rulers to determine the direction from one point to another was examined. This direction is the True heading and, in the absence of extraneous factors, is the course you must sail to keep on a straight line from "A" to "B". Duly corrected for variation and, if necessary, deviation, it becomes the course to steer which you will hand up to the helm, assuming you are motoring in calm conditions or sailing fairly upright, and that there is no current. For much of your sailing, in many waters, the basic course will suffice, but on some occasions it will be necessary to consider factors for which it must be modified.

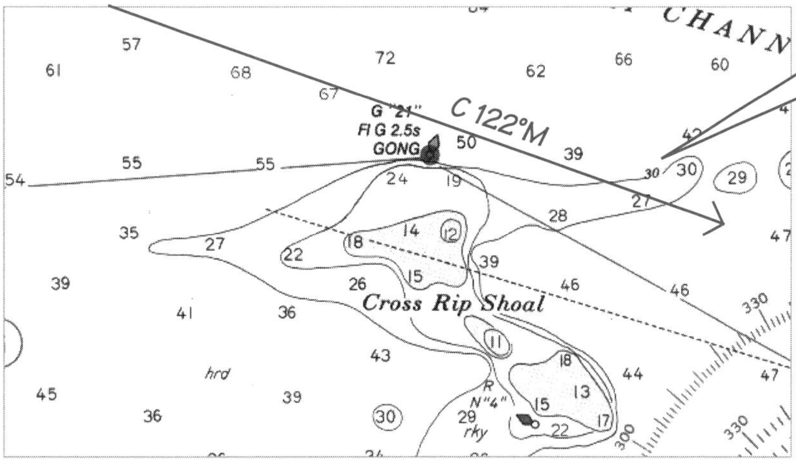

A **course to steer** is denoted by one arrowhead somewhere along its length, so that it cannot be confused with any other line you may draw. Notice that in this case the navigator has written the course to steer along the line, followed by the letter "M" (Magnetic), so there can be no mistake as to whether it is Magnetic or True. This has become common practice among coastwise navigators and is acceptable so long as the passage is short, simple and employs a single navigator always on call. Otherwise, the course is always recorded in the ship's log (see Chapter 9).

Leeway

It is inevitable that when you are sailing close-hauled the boat will slide sideways to a certain degree. This is called *leeway*. It increases with wind and wave state, and decreases as your course direction is further from the wind. By the time the breeze is abaft the beam, it has become insignificant in all but the worst conditions, but never underestimate its effects in a gale. They can be dramatic and unexpected.

Assessing Leeway — Remember that small, unweatherly yachts make far more leeway than large, powerful ones. A visual check can be made by observing your wake as it streams away over the weather quarter. Try to guess the angle it makes with the fore-and-aft line of the boat. The higher the angle, the more leeway you are making. You could even try bringing the chart plotter on deck in calm conditions. Standing in the stern and looking across it at your wake will give you a better sense of leeway angles.

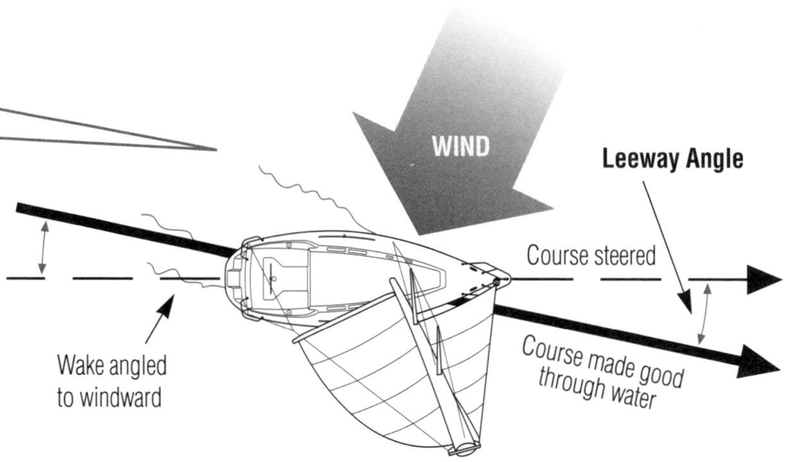

An average 35-foot cruising yacht will make up to 7° leeway close hauled in 20 knots of true wind in open water. Close hauled in 30 knots, it will be lucky to be making less than 10°.

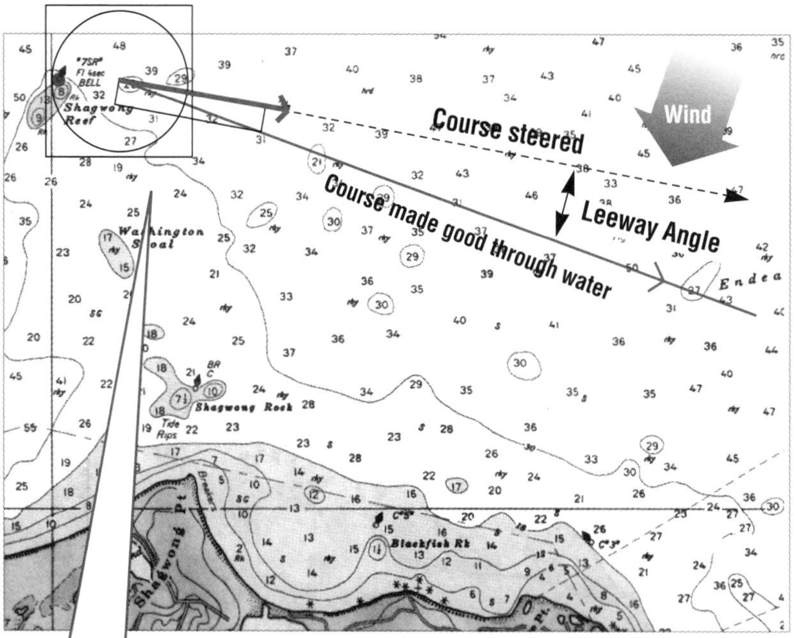

You can counteract leeway by steering to windward of the course you want to make good. To quantify this, place your plotter on the course line and rotate it to windward on the chart by the estimated leeway angle. Draw a short line *(course to steer corrected for leeway)* as shown with an arrow on the end of it. That is the course you should steer, which you will note in the ship's log if you are navigating formally. If you are writing your courses on the chart, you should note the plotted course value rather than the course actually steered, because that is the course you hope to make good after leeway has been taken into account. It doesn't hurt to note both courses in the ship's log.

Sailing in Current

Some parts of the U.S. coastline do not experience significant current, but enough areas are subject to current that everybody should be prepared to deal with it. A yacht setting out from Florida to Bimini or Grand Bahama, for example, must make significant compensation to its course for the powerful Gulf Stream.

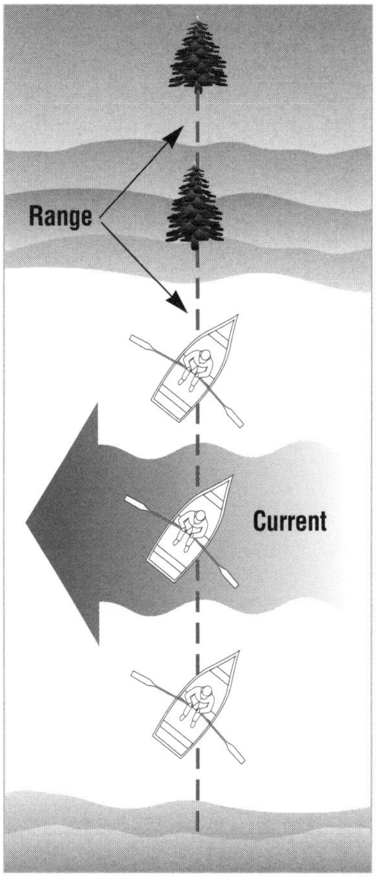

The example at right illustrates the concept by showing a rower crossing a moving stream. The rower eyeballs the destination tree and lines it up unconsciously with a tree behind it. This establishes a *range* (*see Chapters 9 and 12*) which provides a reference for keeping the boat on a straight line across. Because of the sideways motion of the current, the boat needs to steer up into the current to keep on the desired track. It crosses the river on a straight line to its destination, but never points directly to where it wants to go. This technique of establishing and using a range is the best possible way to handle cross-currents when you are within sight of your destination.

Now lets get into a bit of theory, because that is what you will need to understand if you are crossing a current and cannot see your destination.

Suppose you are rowing at 2 knots and the current is running at 1 knot...

▶ First draw in a line joining your point of departure with your destination. It is called your track and is designated by two arrowheads to distinguish it from all other lines drawn on the chart.

▶ *Now suppose you are the rower and you stop rowing for one minute at point "A".* See where the current will carry you to (point "B") and plot a line to show it. This line is called the current vector and is marked by three arrowheads.

▶ *Now imagine that at the end of the minute, the current could be switched off and that you had to plot a course to get you back to the track in one minute.* Obviously, since you are doing 2 knots and the current is doing only 1, you will travel twice the distance that the current drifted you.

▶ Open your dividers to a distance twice the length of the current vector (the actual distance doesn't matter, so long as it is double the current, because this diagram works on proportions), place one point at the end of the current vector (point "B") and scribe across the track with the other point. Where you "hit" the track is point "C".

▶ Join B and C. Mark the line with one arrowhead, and that is the course that would bring you back to your track, the *course to steer*. You could now switch on the current and repeat the operation.

▶ In reality, of course, the current runs all the time. The dynamics are that the boat steers "off" on a parallel heading to the course to steer and is kept on track as surely as it was by the rower eyeballing the original range. Theory meets practice perfectly in this simple vector diagram.

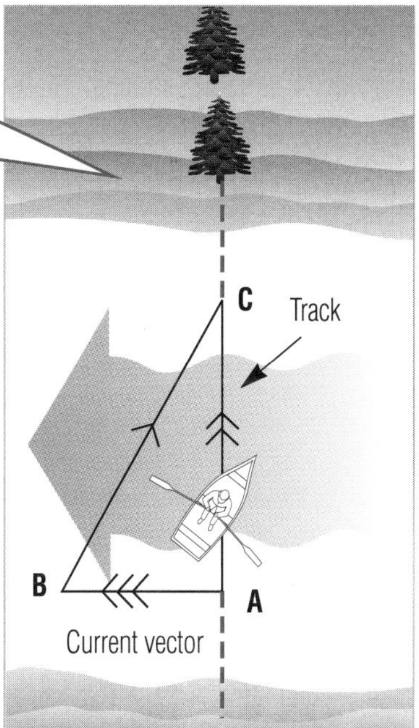

In Practice On the Chart

Deduce from your tidal current chart the probable *set* (direction) and *drift* (speed) of the current for the next hour. Plot the *current vector* (three arrows) from your departure point in the direction of the set. Its length is determined by the drift, so that for a current of one knot the line is one mile long.

Estimate probable boat speed over the next hour and set up that number of miles on your dividers. Place one leg of the dividers on the end of your current vector (B) and scribe a mark on the track with the other (C). Join these two points with a third line. Adjusted if necessary for leeway, this is your course to steer and is denoted by the usual single arrowhead. DO NOT join the end of the current vector with your destination. The course line will fail to coincide with boat speed and the diagram will not work. It doesn't matter if the course to steer does not appear to reach the destination, or joins the track beyond it, because the yacht stays on the ground track all the way. If the passage is more or less than an hour, don't worry. So long as the track passes through your destination, you will arrive.

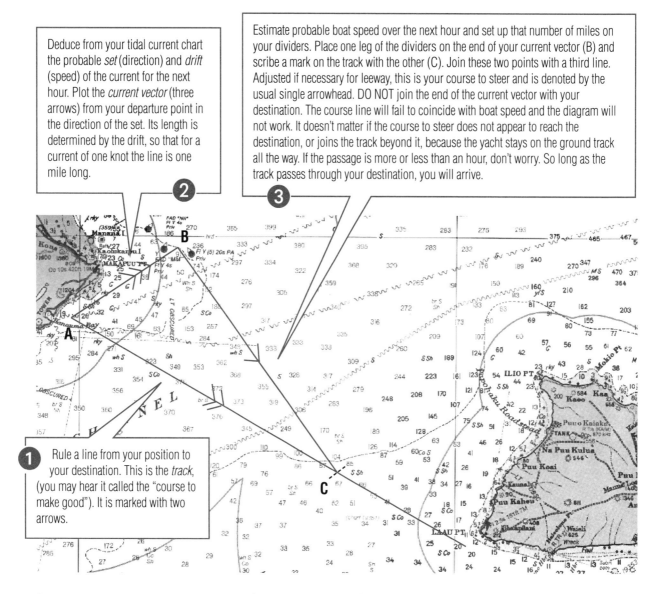

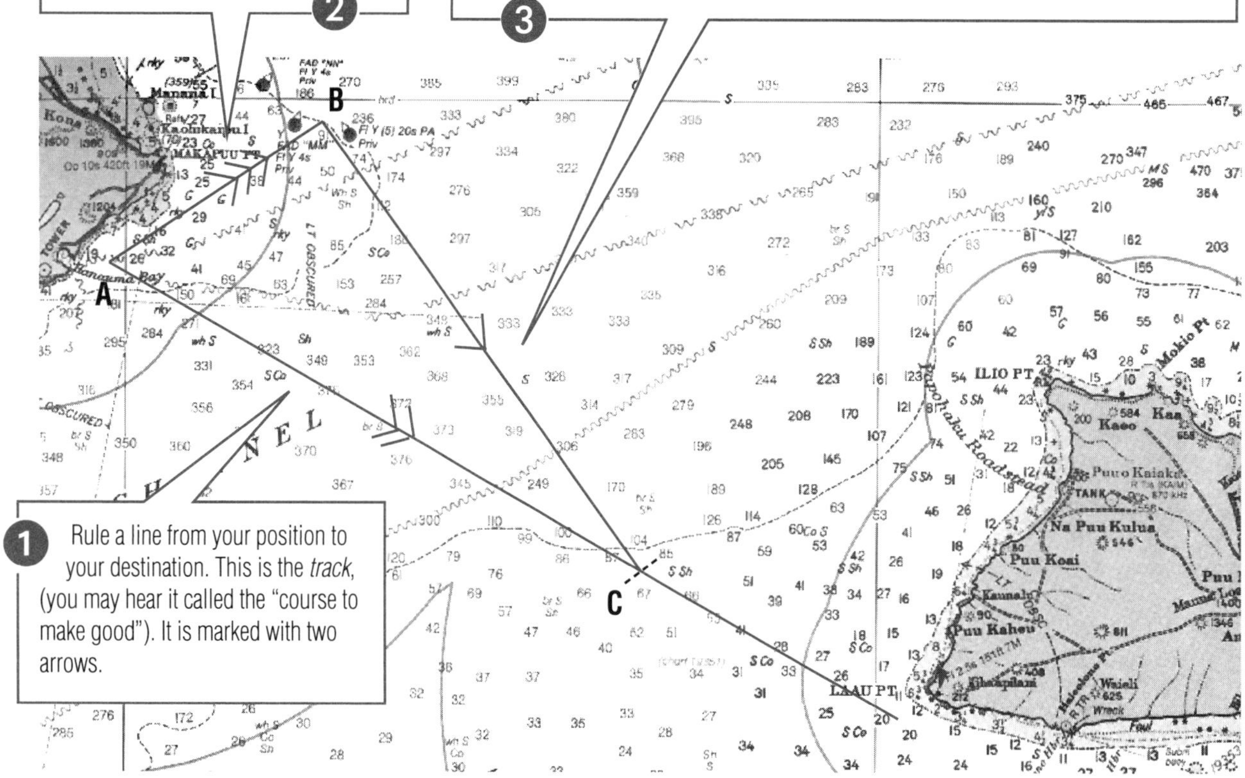

Rule a line from your position to your destination. This is the *track*, (you may hear it called the "course to make good"). It is marked with two arrows.

Remember that this vector diagram is a theoretical representation only. The actual distances are unimportant as long as the proportions are correct. If you record your courses in the log, there is no need to write them on the chart, as long as there is a time noted at the departure point. The time will give you a reference point to find the course in the ship's log and your chartwork will be tidier as a result. If you are not using a ship's log, it is vital that you also enter the log reading on the chart at each known position.

If you are going to build in a leeway correction, do it as described on page 46, then enter the corrected course in the "course" column in the ship's log, because it is from that course that you will build up your Dead Reckoning or Estimated Position. Should you be navigating on the chart alone, write the uncorrected course against the course line, but plot a short line with an arrow (*see page 46*) at a shallow angle to weather of the course to remind you that you have told the helmsman to "steer up 5 degrees for leeway."

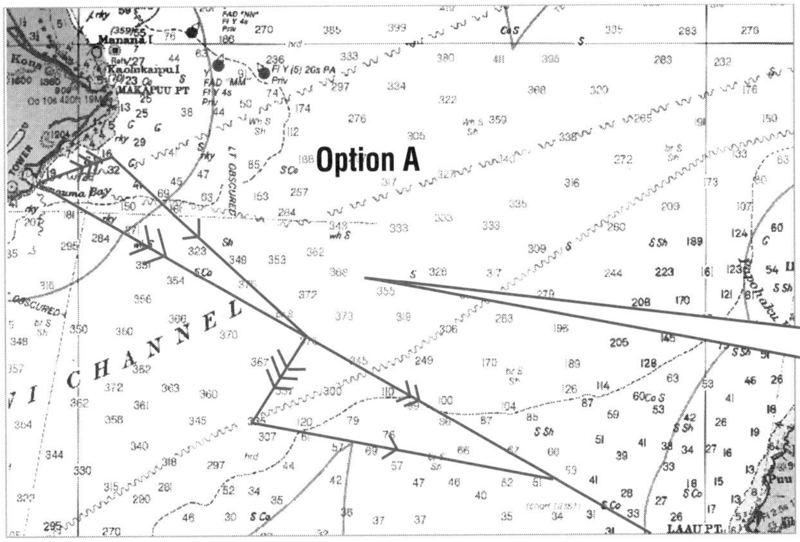

If you are due for a change in the tide, you can enter your tide vectors at the beginning and also at the point where the tide changes (Option A), or you can enter both vectors at the beginning of your plot (Option B). A yacht using option "A" will stay on the track but will travel further through the water in order to do so. The end result of the course steered in "B" is the same as the single one in "A", but "B" progresses further down the track in the same time. A yacht using Option B will therefore arrive in the vicinity of the destination earlier. Note, however, that the yacht in "B" will drift to one side of the track then back again. Usually this will not matter, but be careful of dangers close to the track. If there are any, "A" may be preferred, sticking closer to the track regardless of distance through the water.

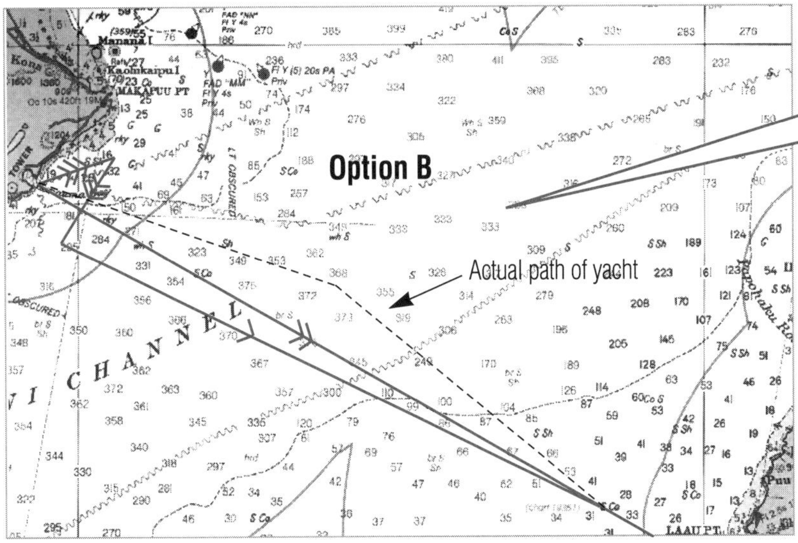

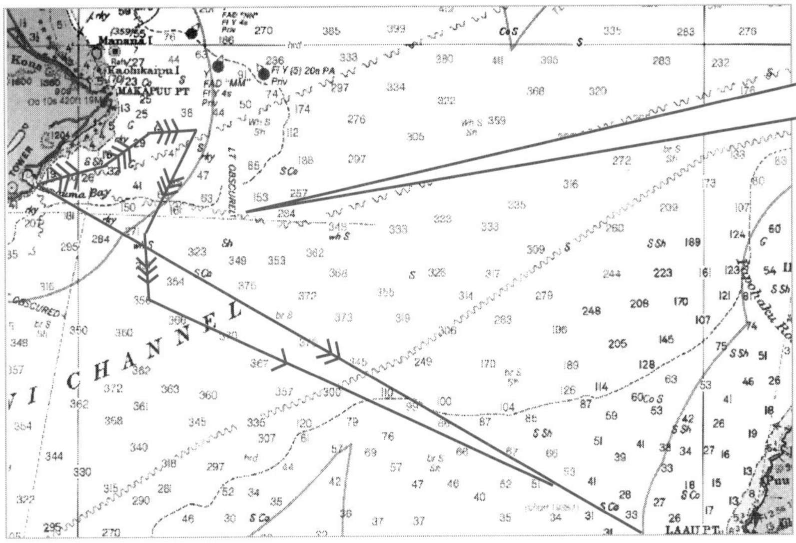

A passage of half a day or more follows the standard rules. If the tides run purely from left to right of your desired track and vice-versa, you may do better to add up all the "left-going" and all the "right-going." Subtract the lesser from the greater and that is your net vector for the anticipated length of the trip. Circumstances may cause you to modify the course to steer as such a passage proceeds (see Chapter 17), but a net vector will give you something useful from which to begin plotting.

Alternative Plotting Method

The plotting arrowheads recommended by US SAILING have been developed from European plotting conventions because they are unambiguous and immediately readable. An alternative method of plotting is to dispense with the arrowheads and to label each line instead. This method is used by American big-ship and commercial navigators. It works well for vessels which are moving in straight lines, which sailing boats often do not, and it is not so easy to read at a glance if you come down to the chart table tired, under stress and perhaps a little seasick. Nonetheless, you may encounter this plotting convention, so you should be aware of its existence.

Course to Steer — This is plotted as a straight line with its direction noted alongside as shown and labeled, "C". The compass direction is conventionally in degrees True and as such is not given a designation. If you choose to plot in degrees Magnetic, it is essential that you label the figures "M". The projected speed of the vessel is also noted in knots along the course line, though sailing vessel navigators using this plotting system sometimes dispense with this. With this system, it is standard practice to measure distance run along the course and plot DR positions as required. This is OK, as long as you have not deviated from the course you set and that there has been no leeway.

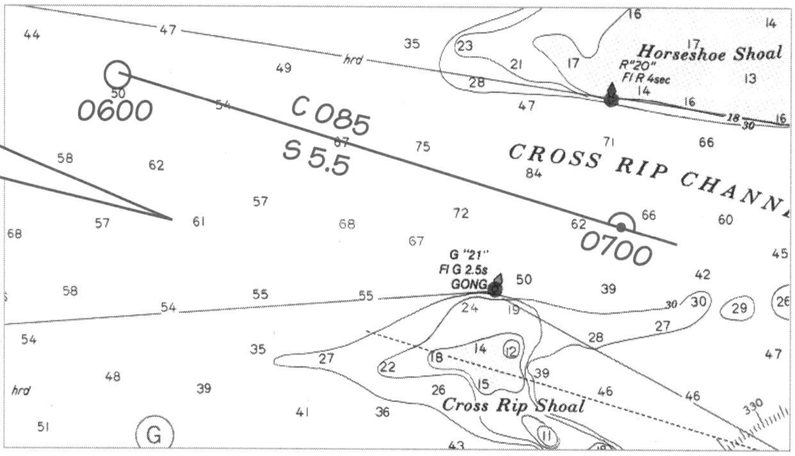

Plotting for Current Effects — To construct a Course to Steer diagram for a crosscurrent, proceed exactly as in the US SAILING System, but do not use arrowheads. Certain writers recommend arrowheads on some lines, but BOWDITCH, the main US authority, does not use them.

Current Vector (Set and Drift) is labeled "S" on one side of the line and "D" on the other. You may write in the figures if you choose.

Course to Steer and "Boat Speed" carry "C" and "S" as shown

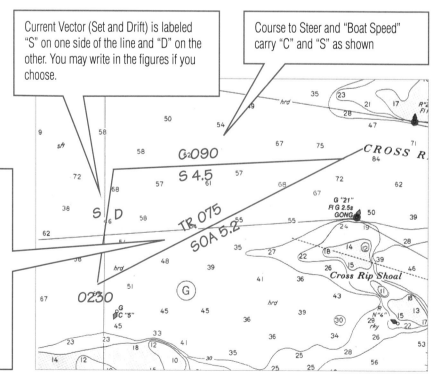

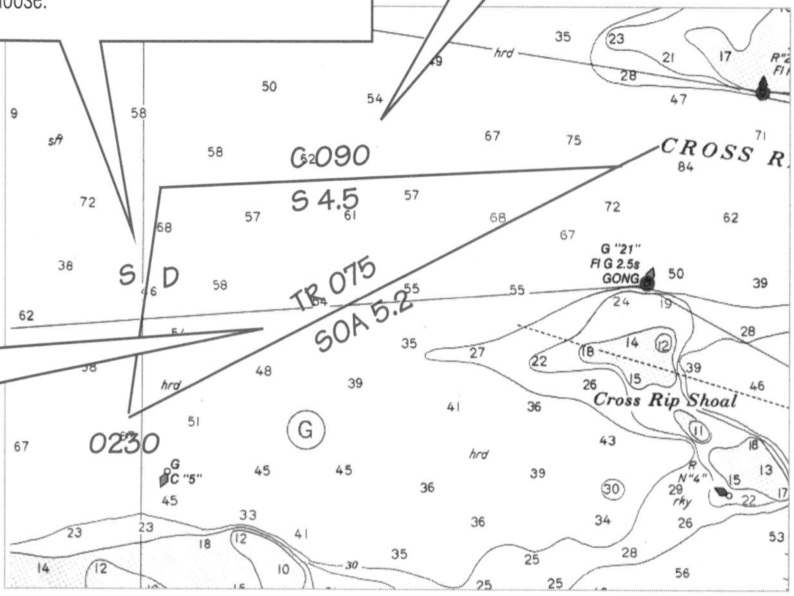

Track is labeled "TR". If you require to know what your speed down the track will be, this is called "Speed of Advance" and is labeled "SOA". You can work it out by measuring the distance along it from the departure point to the position at which the course to steer rejoins it. This is how far you will travel in one hour, assuming your boat sails at a steady speed.

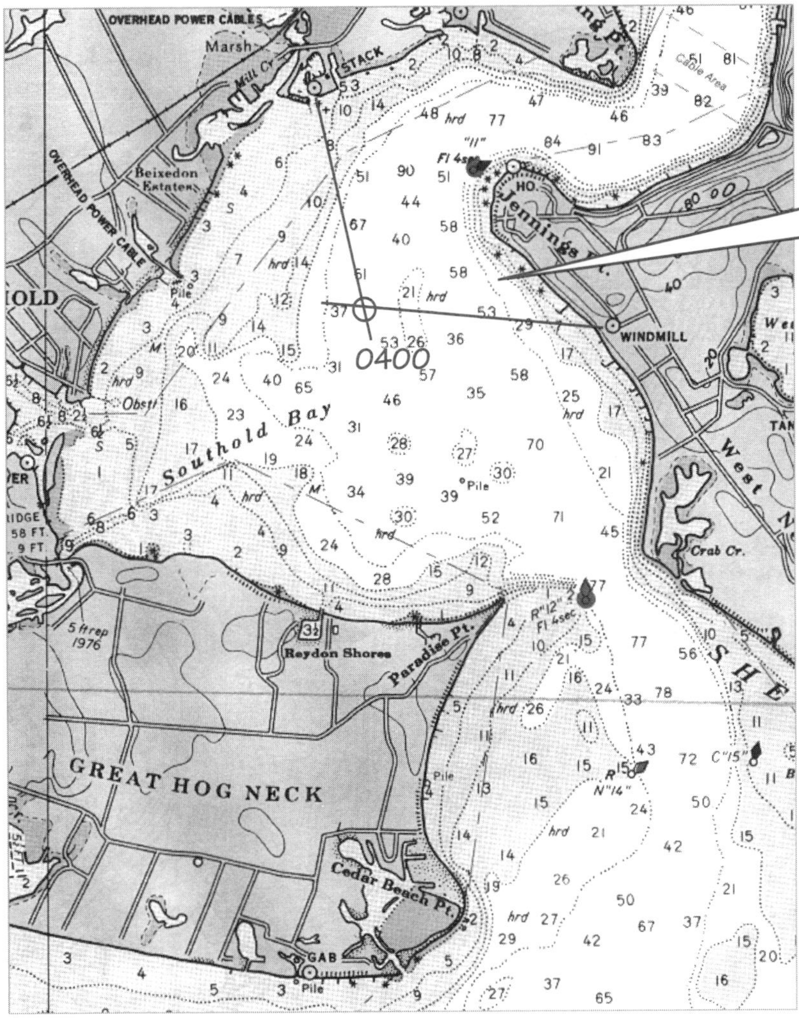

Labeling — Under this alternative convention, *Lines of Position* (LOPs), which are described in Chapter 10, are unlabeled, unless one is taken and plotted singly, in which case its time is noted alongside it. *Fixes* are labeled as described in Chapter 10. If a single LOP is advanced for a running fix, it is labeled first with its original time, then with the time of its advancement (e.g. 1430-1515).

NAV. NOTES...

Now is a good time to review the <u>Plotting Symbols</u> recommended by US SAILING for chart work. More will be introduced later.

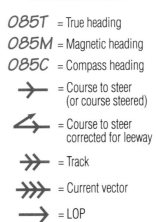

085T = True heading

085M = Magnetic heading

085C = Compass heading

= Course to steer (or course steered)

= Course to steer corrected for leeway

= Track

= Current vector

= LOP

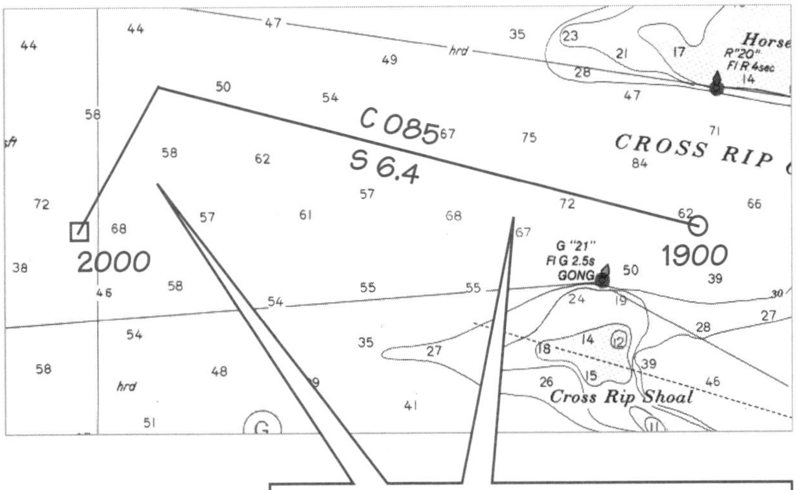

Estimated Position — Course steered and tidal vector lines for *Estimated Positions* (*see Chapter 9*) are plotted without arrows, as shown.

CHAPTER 9 Estimating Your Position

Just as the last chapter looked at one of the two great navigational questions, how to get to where you are going, this one and the next deal with the other, which is working out where you are. These chapters concentrate on finding your position by traditional means; the first by deduction and the second by observation. They show you how to assess the likely accuracy of your estimates and, in doing so, teach you how to think like a navigator. When you have fully understood these you are ready to move on to Chapter 11, Electronic Navigation Aids.

The requirement to navigate by classical, or traditional means, is not set up by a bunch of reactionary fossil navigators who cannot let go of the past. It is a primary safety requirement. If your electrics go down, and voltage is always at risk in a small sailing vessel, you will have to revert to traditional means, even if you do not use them on a daily basis.

But the matter runs deeper than that. Electronics are referred to as Aids to Navigation because that is what they are. They can fix your position with amazing accuracy and give you a pinpoint course to steer, but can you interpret the data you are being given? Can you ask the computer the right questions if circumstances contrive to move the "goal posts" of your navigational requirements? The answer to these vital questions will be a resounding "NO" if you are not a master of traditional navigation. Electronics make a good navigator better, but they may well delude a novice into a false sense of security which leads to downfall.

So be warned. Navigation is not only a means of getting around the seas, it is also a way of thinking, a philosophy. Deciding where you are is half of it, but the remainder is learning how to live with the knowledge that for much of the time you are less than certain of your position. In this chapter we will be dealing with unconfirmed or partially confirmed estimates of position, but first we shall look at the ship's log, or Log Book. Used sensibly, this will provide all the information needed to work up the best available position at any time. The effort involved is minimal.

The Ship's Log

The Basic Form — The ship's log is used to note any incident or item of navigational significance. There are always discussions about how often entries should be made if nothing in particular is occurring, but in addition to "natural entries" such as course changes, passing a buoy, or being headed off your course, "every hour on the hour" is a good rule of thumb.

A ship's log need consist of nothing more than a simple exercise book with its pages ruled off into a few columns. Entering items of navigational interest becomes a second's work and the knowledge is stored permanently in case you need to refer back to it, which in many cases, you will.

TIME	LOG	COURSE	WEATHER	REMARKS	ENGINE
0600	*0*	*176 C*	*N10 1005 mb. Fair*	*Weighed anchor from Camden Harbor.*	*30 min.*
0630	*2.0*	*194 C*	*N5 1005 mb. Cloudy*	*G7 buoy at hand. Alter course to 194 C.*	
0700	*4.3*	*194 C*	"	*Fix taken on Lowell Rock light, Glen Cove Tower and Graves Buoy. Fix on chart.*	
0750	*8.2*	*163 C*	"	*G3 buoy at hand. Alter course to 163 C.*	
0800	*8.9*	*163 C*	*N5 1004 mb. Fog*	*Display radar reflector.*	
0900	*13.3*	*163 C*	*N5 1003 mb. Fog*	*EP and GPS fix on chart.*	

Time: The time of the incident, usually expressed in local time

Log: The reading of the ship's distance log

Course: The actual course being steered, or the new course if an alteration is ordered. If the course steered has varied from the course asked for, then record the course you actually managed in the next log entry, because that is the one you will be using for estimating your position.

Weather: Record the wind and anything else interesting. Barometer reading can be a lifesaver if the instrument is doing handstands. Use your own abbreviations to keep it simple, e.g. "SW10 1003."

Remarks: This column is to note what it is you are recording. Typical examples are, *"Fix on chart"* (see Chapter 10), *"EP on chart"* (see below), *"Skulking Cove buoy 50 yards to port," "Visibility seriously deteriorating," "Entering Eastbound shipping lane."* On a lighter note, you might care to add, *"Breakfast a nominal affair owing to 15 foot seas breaking over deck."*

Engine: On many boats it is valuable to have an idea of your engine hours. These can be recorded here (*"on"* / *"off"*). You can also record any other factors of interest (*"running 20 degrees warm"*), periods of topping up sump oil, etc.

Chuck Place photo

The example below is all you really need for a ship's log. If you want more, go for it, but the format here will see you through.

The ship's log is also required in the unlikely event of any "official" interest in your passages. It is helpful to be able to show where you have been, and when you were there. In addition, the book forms a diverting record for personal reminiscence because, in addition to impersonal data, a good yacht log contains comments about the quality of the lunch served up at the height of the recent gale, what the mate said when the skipper was late on watch, the amazing arrival of Wallace Whale at 1245 in 43°N 68°W, etc. This type of entry does no harm at all, and can encourage otherwise recalcitrant watch keepers to do the honest thing and record that buoy ("Now, what was its number?") they passed in the fog at 0614.

As we have noted, some navigators prefer to write their navigational information on the chart "as they go along." Within the constraints mentioned, this is OK, but you must plot every significant thing as it happens, making sure that your figures and letters cannot possibly be mistaken for anything they are not. In a strange area or on any but the simplest passage, this can make for some very fussy chartwork. An uncomplicated ship's log is therefore recommended.

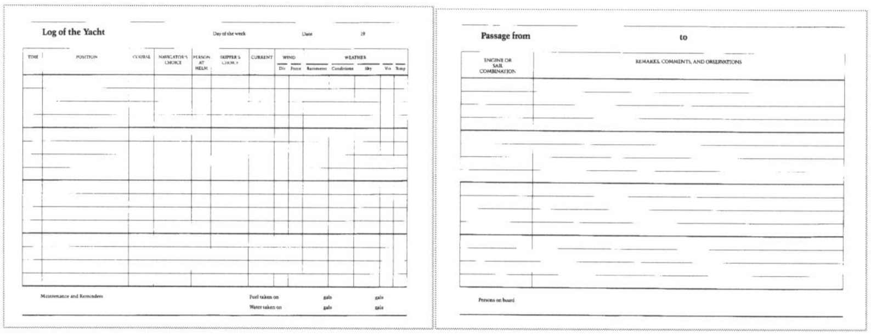
Log of the Yacht　Day of the week　Date　19

TIME	POSITION	COURSE	NAVIGATOR'S CHOICE	PERSON AT HELM	SKIPPER'S CHOICE	CURRENT	WIND		WEATHER				
							Dir	Force	Barometer	Conditions	Sky	Vis	Temp

Maintenance and Reminders　Fuel taken on　gals　gals
Water taken on　gals　gals

Passage from　to

ENGINE OR SAIL COMBINATION	REMARKS, COMMENTS, AND OBSERVATIONS

Persons on board

Mystic Seaport Museum Stores

The sophisticated ship's log — For "professional" yachts, a more complete ship's log form is often called for, actively encouraging the watch to make detailed entries every hour on the hour. Such logs are more like those used by large vessels and are a sensible precaution for yachts in commercial or potentially commercial situations. This ship's log is the one used by the schooner *Brilliant* from Mystic Seaport. It is a model of comprehension and was worked up with the cooperation of the late Rod Stephens.

Deducing Where You Are

If you know your position when you start (your departure point), and you know how far you have sailed and in which direction, you can use this information to work out where you are now. A position plotted by using these two basic inputs is called a *Dead Reckoning* (DR) position. It is worked up from Distance Run (read from the log) and Course Steered (as noted from the compass).

If your log read zero at the departure point, what it reads now is how far you have come. If it started at 103 miles and now it says 110, your distance run since taking your departure is 110 - 103 = 7 miles.

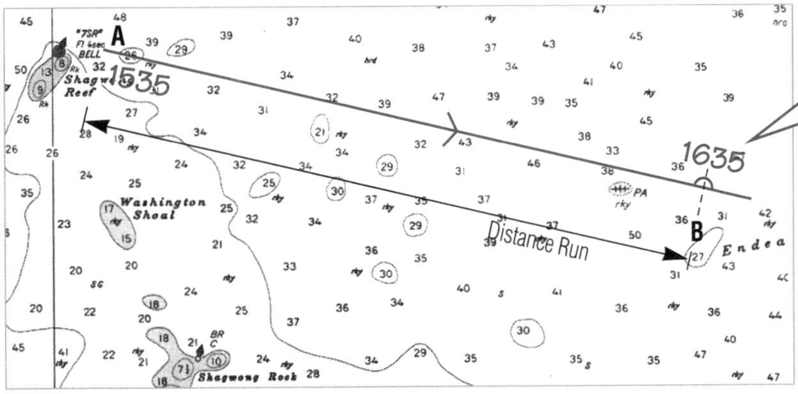

To Plot a DR Position

To plot a DR position, you must start from a known position which may be a fix, an Estimated Position or close observation of a known feature, etc. Next, rule a line representing the course steered. Set up your dividers to your distance run, then place one point at your departure point (A) and scribe across the course line with the other (B). Mark as shown and that is your DR position. If it is not to be further refined, record it, either in the book or by noting the time and log reading on the chart. *Note the plotting symbol for DR position, the times written at each end, and the single arrowhead on the course line.*

The Estimated Position

If there is any leeway or current to take into consideration, a DR position is not going to be accurate enough for serious navigation, because the course steered and even the distance run may have been "got at" by these extra factors. In these circumstances, a DR is refined into what is called an Estimated Position (EP). An estimated position is therefore a DR position corrected as far as possible for current and leeway. It is the "best estimate" that can be achieved without involving data gathered from sources external to the yacht.

To Plot an EP

Start from a known position, then plot a line representing your course and distance made good through the water. This will be either the course drawn on the chart (if you are not using a ship's log), or the course actually steered, corrected for leeway by rotating the plotter to leeward by the required number of degrees as shown. Conventionally, it will represent an hour's run. Mark the line with a single arrow.

1

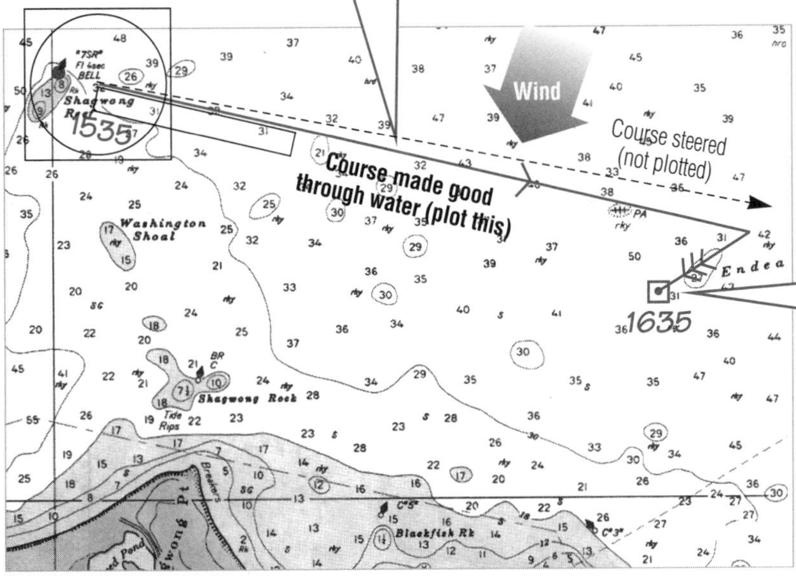

2

From the tidal current chart, your observation of the current, or current tables, decide the current's direction (set) and strength (drift). Plot an hour's set and drift from the end of your distance run line and mark the line — the tide vector — with three arrows as usual. The end of this line is your Estimated Position. Mark it with a dot enclosed by a square, write the time against it and record it. If you are not using a ship's log, you must write the log reading against the EP as well as the time. If you do not do this, you might as well have saved your pencil lead.

If you require an estimate of your position at some time which does not coincide with a "tidal hour", you must do a little mental arithmetic to determine the length of the tidal vector. Suppose, for example, you run just 40 minutes. You will divide the hourly rate by $\frac{2}{3}$ (40 minutes being $\frac{2}{3}$ of 1 hour) for the purposes of your plot.

EPs Plotted at Odd Times — It is not always convenient to work up an EP conveniently timed to coincide with the tidal hour. Nor is it always necessary to plot an EP every hour.

NAV. NOTES...

At the end of the last chapter we reviewed some plotting symbols recommended by US SAILING for chart work. Following are some *<u>additional plotting symbols</u>*, *for more advanced piloting.*

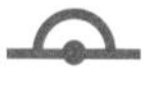 = DR position

 = Estimated position

 = "Conventional" fix

 = Electronic fix

Plotting an EP After a Number of Course Changes

If you require an EP after a number of course changes, such as during a long beat to windward where the yacht tacks several times, there is no need to plot an EP every time you go about. You most probably will be tacking on advantageous wind shifts which rarely arrive at convenient intervals, so the answer is to work up a series of DR positions (corrected for leeway), then plot all the tide vectors onto the last corrected DR and mark the EP at the end of it.

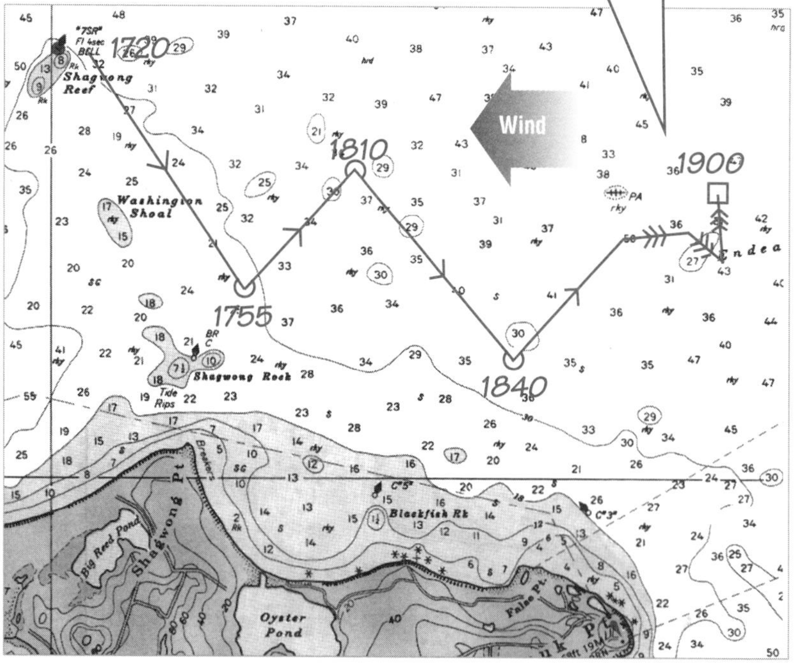

Evaluating an EP — An EP is only as good as the data from which it is plotted, and it must be clearly understood that tidal data are only a prediction. If there is no current, your EP is likely to be more accurate, being then only a DR corrected for leeway.

You should know the accuracy of your log and compass. If their readings require correction, you will need to apply deviation and the log error correction (*see Chapter 5*). A corrected reading from a log of predicted inaccuracy is as good as a reading from a perfect instrument.

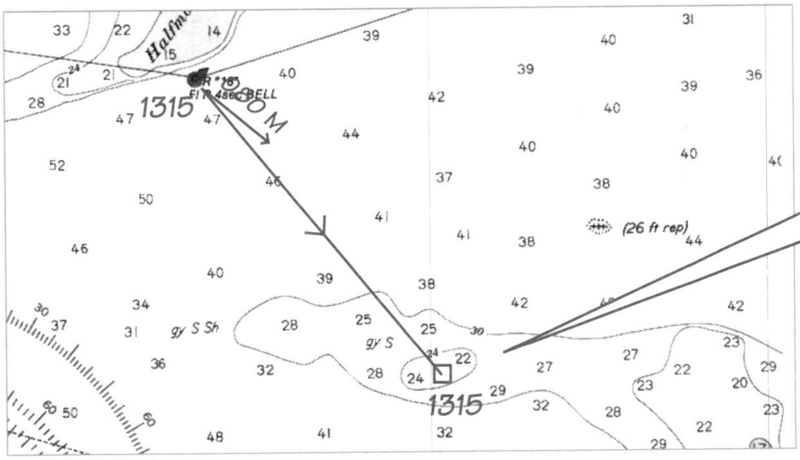

If you must rely on your EP because you cannot manage a full fix (see next chapter), you can sometimes go some way to confirming it by plotting a single line of position (LOP) at the same time. This may be a magnetic bearing of a shore object or buoy *(see Chapter 10)*, or simply an echo sounding. In this example, a depth sounding of 22 fathoms builds confidence in the EP (DR + leeway).

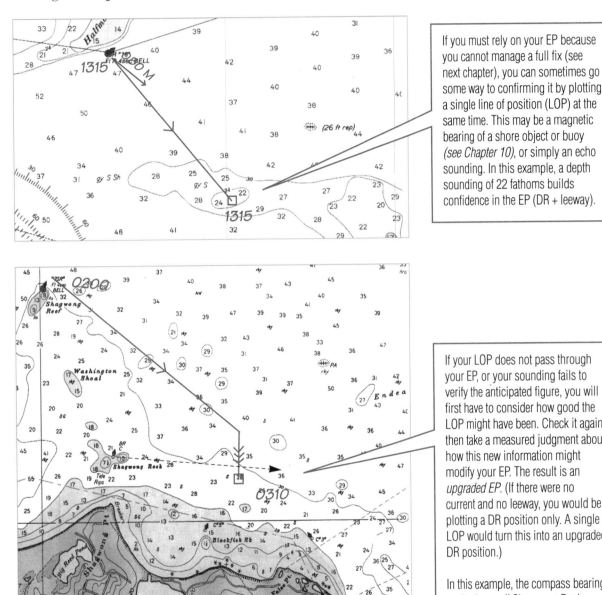

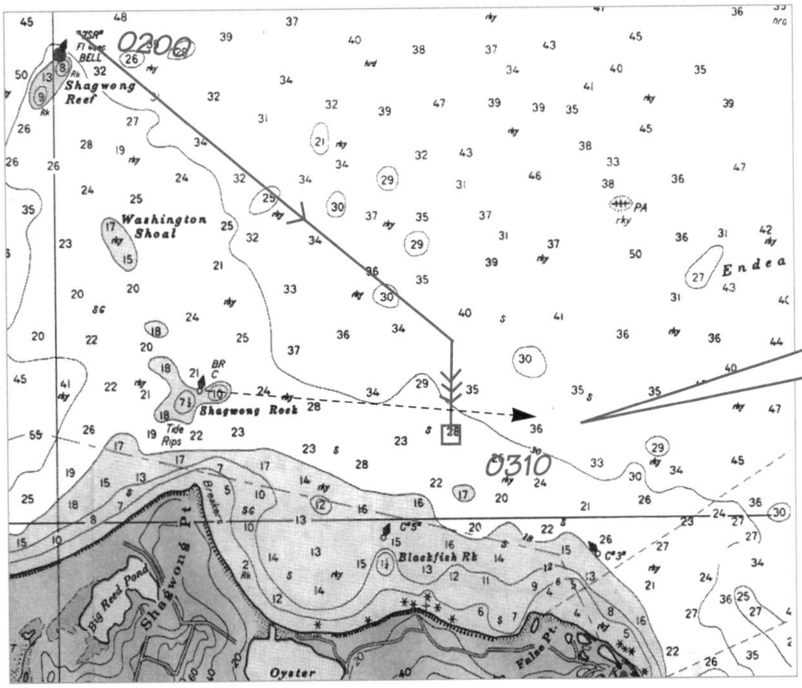

If your LOP does not pass through your EP, or your sounding fails to verify the anticipated figure, you will first have to consider how good the LOP might have been. Check it again, then take a measured judgment about how this new information might modify your EP. The result is an *upgraded EP*. (If there were no current and no leeway, you would be plotting a DR position only. A single LOP would turn this into an upgraded DR position.)

In this example, the compass bearing on the buoy off Shagwong Rock passes behind the EP. Either upgrade the EP or retake the bearing. Remember, however, as you take new bearings, the boat will have moved.

Some navigators call an upgraded DR an Estimated Position, but an upgraded Estimated Position or DR is essentially different from a pure EP or DR because it makes use of information drawn from outside the yacht. A fix, which is a position of considerable accuracy, draws all its data from outside observations, so an upgraded EP or DR is really halfway to a fix. To go all the way, read the next chapter.

CHAPTER 10 Knowing Where You Are

Next to plotting your exact location when you are tied to a dock, the most accurate position is a *fix*. There are a number of ways of achieving a fix, but the basic principle is that you cross check one piece of information against another. It is an axiom of navigational thinking never to accept a single, unconfirmed item of data without searching for a way of corroborating or rejecting it. Occasionally, this will prove impossible. If so, the data must be treated with the gravest caution.

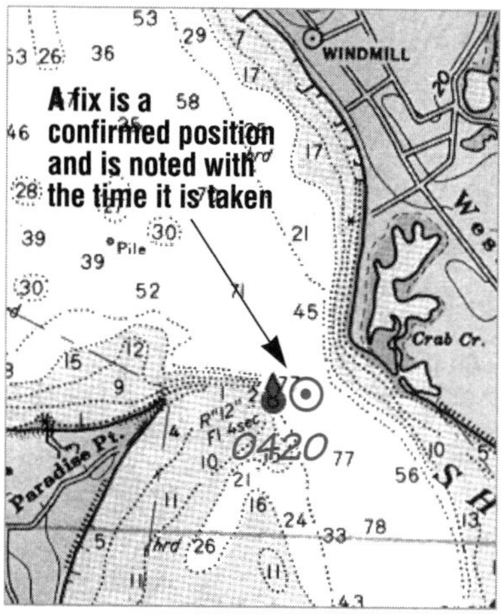

Position by Immediate Observation

If your boat is alongside a navigation mark such as a buoy, you know where you are (assuming the buoy is on station and you have identified it correctly). Your position is said to be *fixed*. It is plotted on the chart by way of a circle around a dot, and has the time noted close to it. It is also recorded, either in the ship's log, or by noting the log reading against the fix.

Positions Defined by Lines

If you are an estimated 50 yards away from the buoy, under most circumstances it will suffice to say, "*I am 50 yards south of Narrow Gut Buoy*" and plot your position accordingly. In open waters you could still treat this as a fix, but as distance from the buoy increases, the value of such a position decreases, depending upon how accurate you need to be.

As you move further away from a known charted object, your estimates of distance from it will become inadequate for fixing purposes. However, if you take a bearing on it using your hand bearing compass and plot that bearing on the chart, you know that you are somewhere on the line. Such a bearing is called a *line of position* (LOP). It is drawn with an arrowhead on its extreme end away from the object and is only as long as it needs to be to pass through your probable position. A single LOP is always plotted with the time that it was taken.

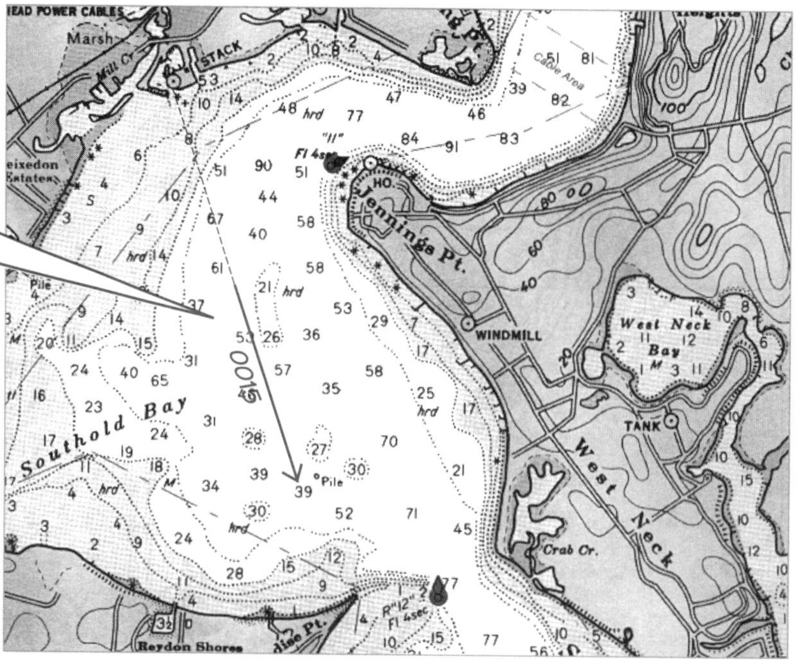

NOTE: Dotted lines shown in illustrations are not normally drawn, but are included for reference purposes.

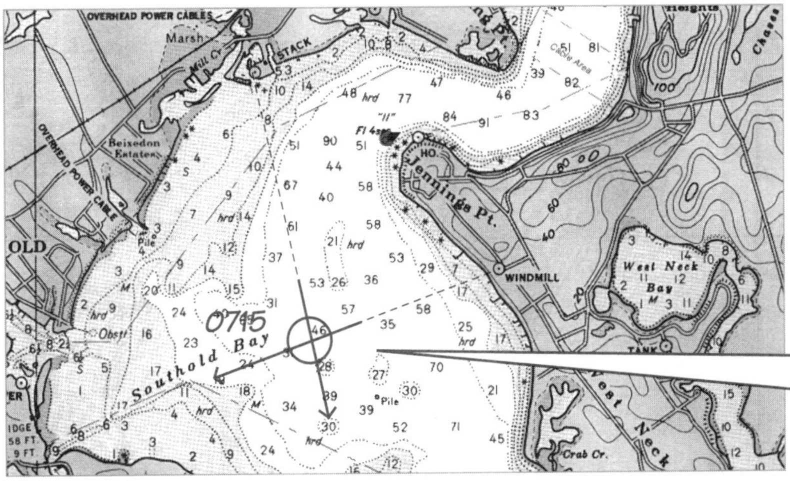

The next step is to find out where on that first LOP you are. You can achieve this by finding another suitable charted object, taking a bearing on it, and plotting its LOP. Where the two LOPs cross, in theory, is a fix on your position.

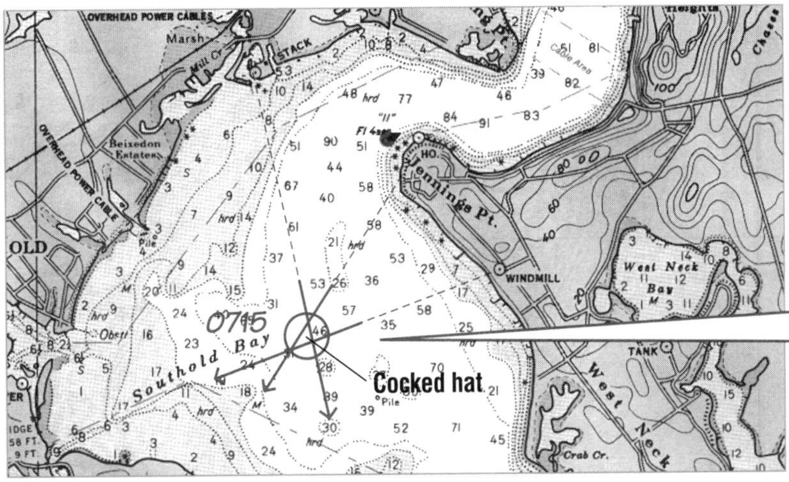

It is always a good idea to check the accuracy of a "two-point" fix because it is probable that at least one of the LOPs won't be as good as you'd like. The solution is to plot the LOP of a third object. This might cross the other two "right on," in which case your fix is confirmed. It is more likely that it will create a triangle known as a *cocked hat.* The size of the triangle will give you an indication of how good your fix is (the smaller, the better).

You now have a classic *three-point fix*. Try to select objects whose LOPs will intersect at 45° or more. This way you will minimize the effects of any errors in the bearings. A shallow *angle of intersection* can lead to big distortions in your apparent position. Objects on either side of you almost opposite one another, for example, give a very poor angle of intersection, whereas an object ahead, another abeam and a third 45° on the bow would be ideal (*see below*).

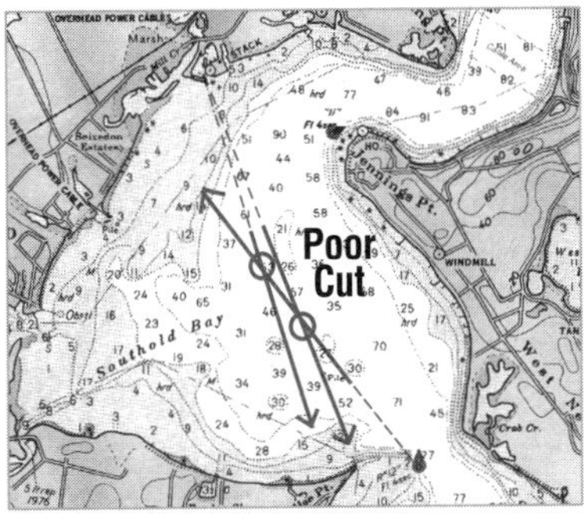

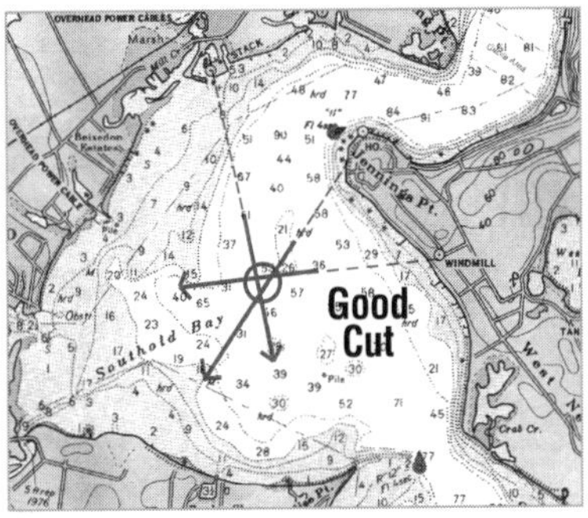

NAV. NOTES...

US SAILING convention draws LOPs with an arrowhead on its extreme end away from the object and is only as long as it needs to be to pass through your probable position. The selected objects, time and log reading should be recorded in the ship's log.

Convention assumes the fix is in the middle of the cocked hat unless there is a danger nearby, in which case assume the fix is closer to the danger.

Always use a closer rather than a more distant object for an LOP when there is a choice. A small error in bearing is magnified with distance. In this example, the error in bearing is the same for buoy A and B, but the error is multiplied over a greater distance to B.

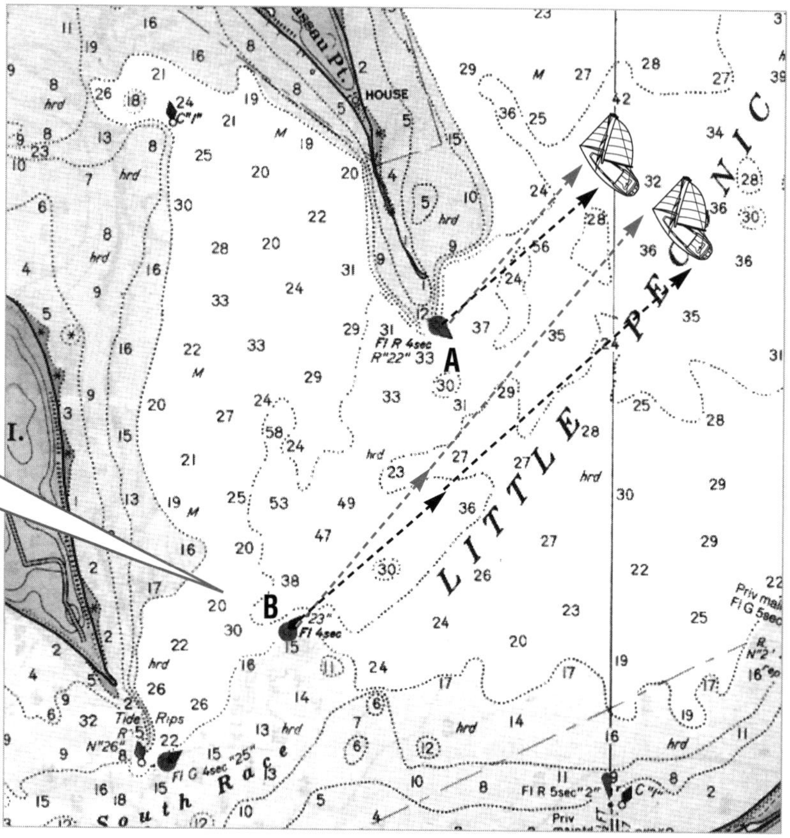

An "official range" set up ashore with two stationary markers provides a great LOP. *Quality is the best.*

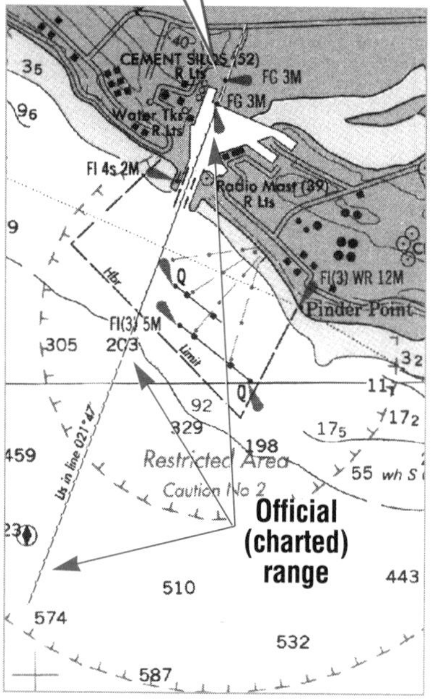

Sources of LOP

When you are looking for a fix, there are many more ways to find LOPs than to grab the hand bearing compass. They are not all noted here, and any navigator can come up with his or her own solutions on any given day. As long as the object is on the chart and can be seen on the water, you can find an answer.

Ranges — A range (sometimes referred to as a *transit*) consists of two objects lined up on the same bearing. It is always better than a compass bearing of a single object because any line joining two known items is precisely defined. You can measure the bearing on the chart, then check it with your compass to make certain you are looking at the right thing when the two objects come in line.

A compass bearing is, as we have seen, always subject to errors. Even if you have found a place on deck where your hand bearing compass seems clear of deviation, try using the instrument in an 8-foot sea while the reading swings 15° either side of what you think you want. You'll soon understand the benefits of ranges.

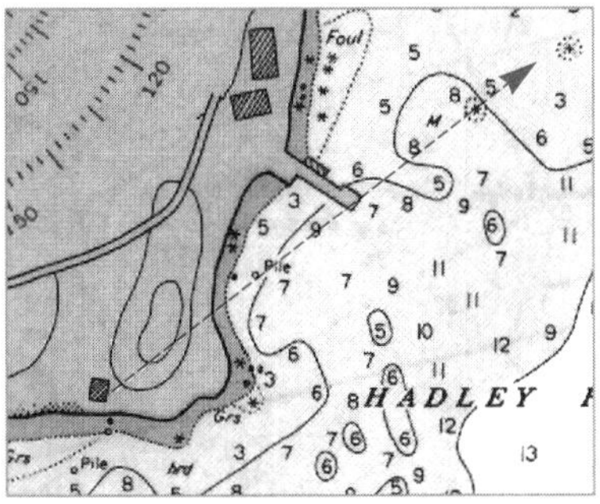

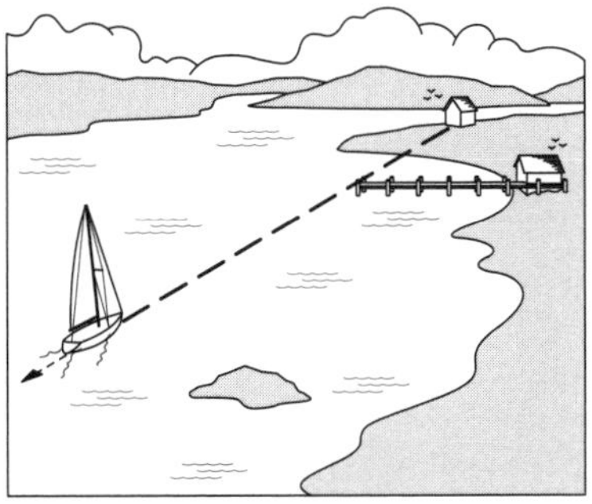

An "unofficial" charted range. This is found on the chart by you, noting two prominent objects whose range will form a useful LOP. The range will not be printed on the chart. A good way to spot such a range is to lay your plotter edge in the approximate direction that you want to find an LOP and search along it for likely candidates. For example, a church steeple "on" with the end of a prominent pier. *Quality is almost as good as an "official" range.*

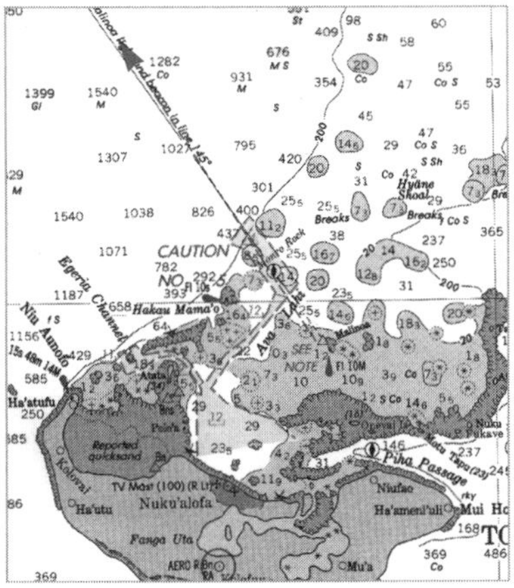

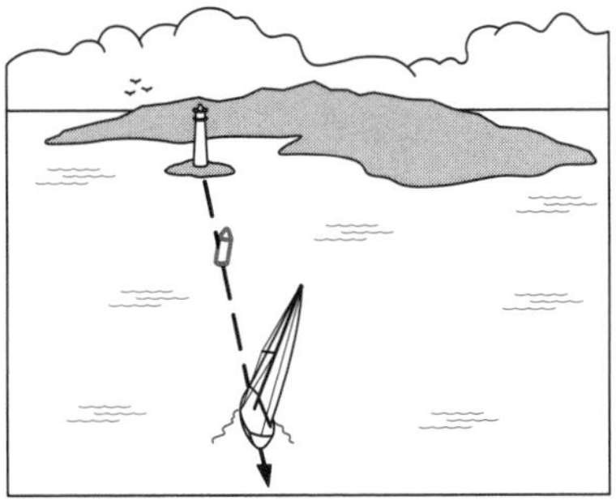

A fixed object ranged with a charted buoy. *Quality is good, but only as good as the stationing of the buoy.*

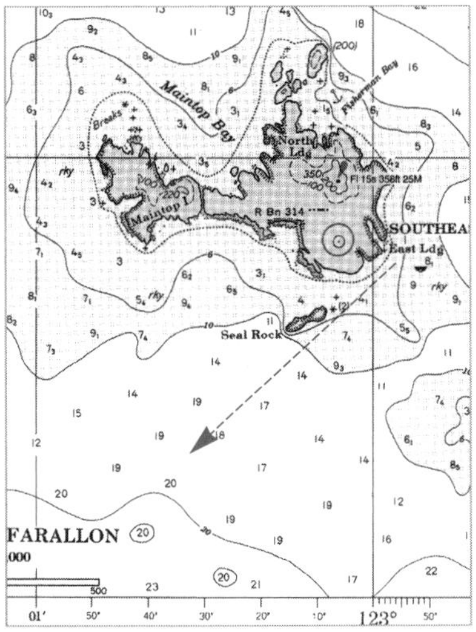

Two headlands in line. *Quality is good*, but there may be some ambiguity about where a headland actually enters the sea, particularly in regions of large tides. Read your chart carefully and expect a decent result, but not perfection. Try to select headlands that meet the water with a distinct edge.

The colored sector cutoffs in lighthouses, or extra lights shining out over a specific sector to mark a danger or a safe passage. *Quality can be extremely accurate if the cutoff is sharp*, but watch the edge of the sector carefully, because they are not all as crisp as you might suppose.

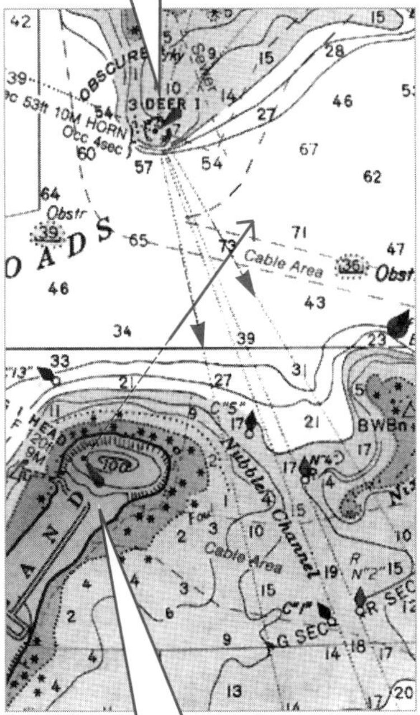

A compass bearing of a fixed object such as a lighthouse. *Quality is as good as your compass, conditions and distance allow.* You will know when you take the bearing if it can be relied upon for tight work or not.

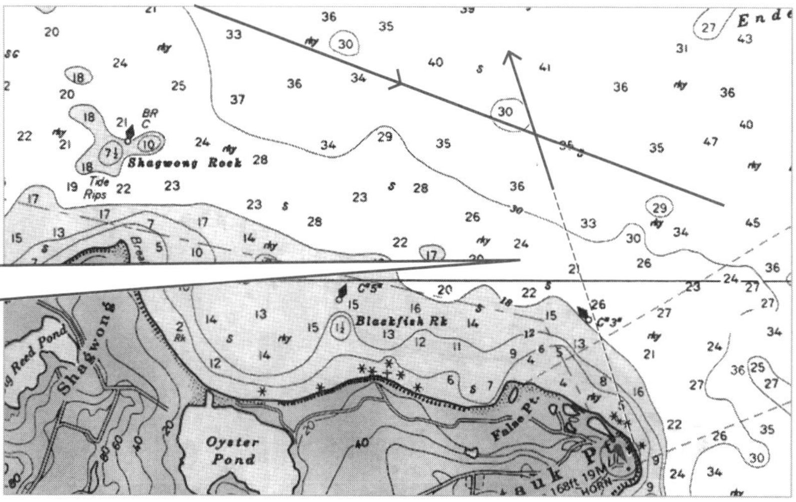

Bearings of topographical features such as hilltops, headlands, ends of islands (usually expressed as LHS [left-hand side] or RHS [right-hand side]), etc. *Quality varies.* These can be good, bad or indifferent. Rely on common sense to decide.

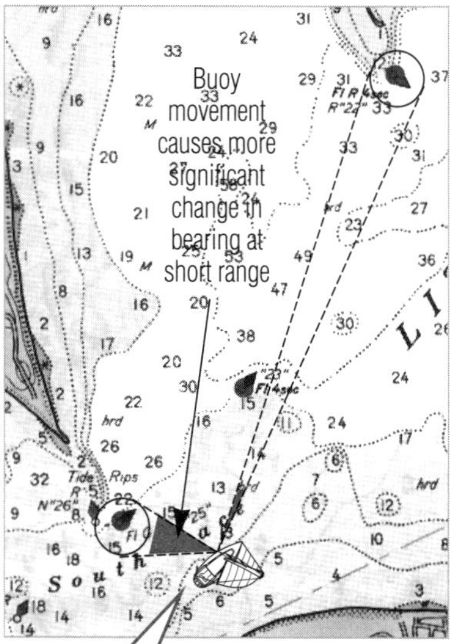

Bearing of a Buoy. Buoys inevitably move around on their moorings, some more than others, so keep this in mind when taking LOPs from buoys. *A fixed object is preferred if you can find one.* Do not disregard the buoy, but be aware of its potential shortcomings, especially at close range where a 25 yard movement from its charted position could be significant. At three miles distant, any such drift becomes irrelevant. If the buoy is seriously off station, Notices to Mariners will have told you, or if not, local Navigation Warnings on the Radio, Navtex, or Harbor Office notice boards.

LOP from an Echo Sounding — If a line of soundings or a depth contour is clearly defined, a sounding may be very useful. If the bottom is as flat as a pool table, it will be a waste of time.

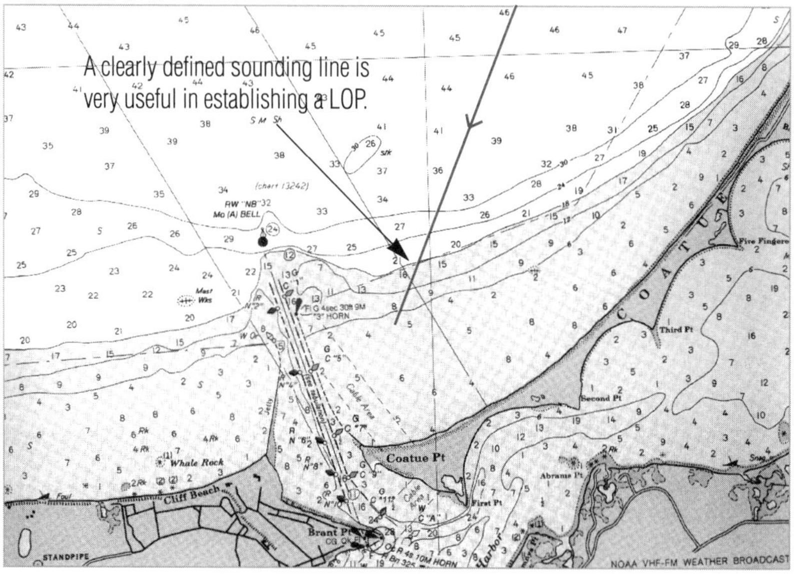

It is a good habit always to check a fix, a DR or an EP against a sounding taken at the same time. You have then put in a further and different source of data and have carried out good navigational practice. If the depth tends to confirm the position you have plotted, so much the better. If the depth seems wrong, you must double check your position to find the wild card. It may turn out to be the depth, but it is just as likely to be that third bearing you weren't really sure about and which you "made to fit." This desire to have a second opinion on everything is the essence of safe navigation.

LOPs by "Distance Off" — If you can see an object and are able to determine your distance from it, you are situated somewhere on a circle drawn around the object with its radius the same as your *distance off*. Such a circle, or a segment of it, forms what is known as a *Circular LOP*.

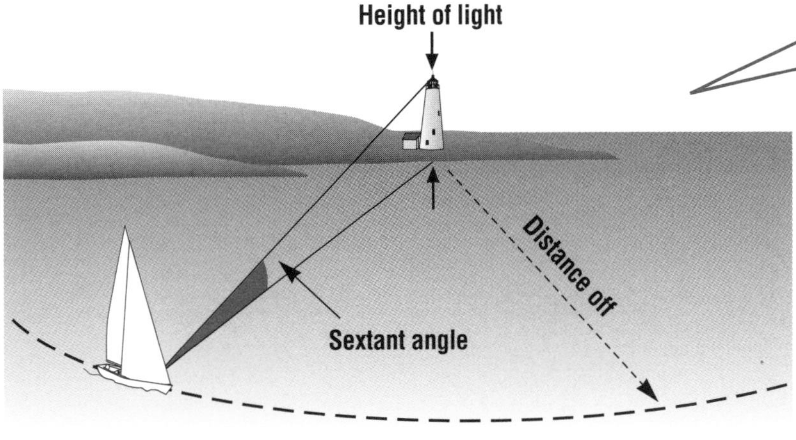

If you have a sextant on board, you can measure the height of the object in degrees and minutes then, using its height as stated on the chart, use a table usually found in your almanac (or refer to the table in the Appendix for determining "Distance off by Vertical Sextant Angle." An excerpt is shown below.) Enter one side of the table with the angle you have measured and the other with the height of the light taken from the chart. The figure where the two intersect is your distance off in nautical miles and cables (one cable = 1/10 of a nautical mile).

TABLE FOR FINDING DIS

Distance in Miles & Cables	HEIGHT OF OBJE			
	12 40	15 50	18 60	21 70
m c	° ′	° ′	° ′	° ′
0 1	3 46	4 42	5 38	6 34
0 2	1 53	2 21	2 49	3 18
0 3	1 15	1 34	1 53	2 12
0 4	0 57	1 11	1 25	1 39
0 5	0 45	0 57	1 08	1 19
0 6	0 38	0 47	0 57	1 06
0 7	0 32	0 40	0 48	0 57
0 8	0 28	0 35	0 42	0 49
0 9	0 25	0 31	0 38	0 44
1 0	0 23	0 28	0 34	0 40
1 1	0 21	0 26	0 31	0 36
1 2	0 19	0 24	0 28	0 33
1 3	0 17	0 22	0 26	0 30
1 4	0 16	0 20	0 24	0 28
	0 15	0 19	0 23	0 26

Reed's Nautical Almanac

There is another table in the almanac (also see Appendix) which gives "Heights of Lights Rising and Dipping." Enter this with your "Height of Eye" (how high above the sea your eyes are) and the height of the lighthouse. The result is the maximum distance you will see the light before it dips or rises at the horizon as you sail away from it or towards it at night.

Fixes from Single Objects

You can achieve a useful fix by a single good quality magnetic bearing crossed with a circular LOP taken from your Distance Off the same object. Such fixes are handy when you see a single lighthouse during the day, or when making a landfall or departure at night. In the first case, a sextant angle is needed, but for the second, the landfall fix, it is enough to note when a light first appears on a clear night (you may see the loom well before the lantern itself shows - it is the lantern you want).

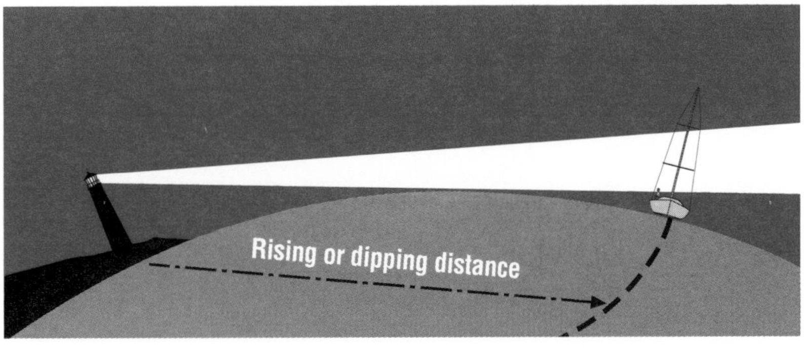

For a landfall fix, take a bearing from the lighthouse and a log reading and note the time the light dips or rises. All that remains is to plot your fix and check the depth, if appropriate.

Running Fixes

Where only one object is available to determine your position, and you cannot work out a distance off from the object, your only recourse is a *running fix*. If there is any current, these fixes are of limited reliability. Even with no current, a running fix is the bluntest of navigational tools. Nonetheless, it must be considered because occasionally it will be all you can manage. As long as you do not expect too much from it, a running fix can help confirm or question an EP, and is better than no fix at all.

NAV. NOTES...

Here are two new plotting symbols to add to the list:

= **Circular LOP**

= **Transferred LOP**

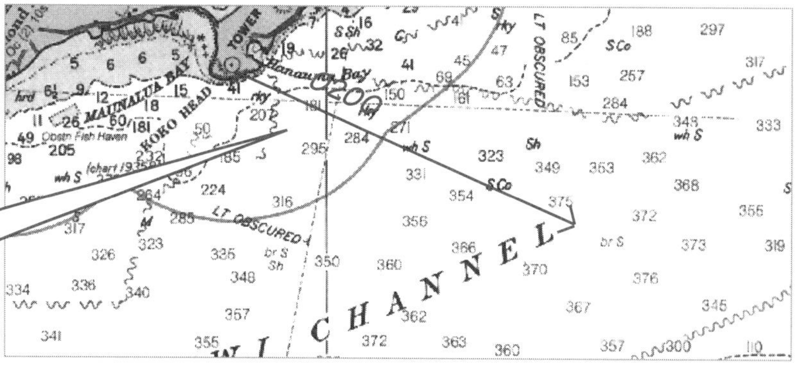

1 First, take and plot a single LOP from an object. Record this and write a time against it as shown and note the log reading.

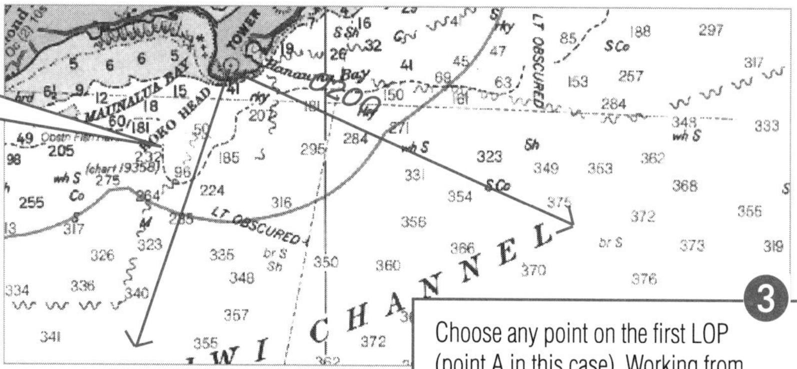

2 Run enough distance to open a good "angle of intersection," then take and plot a second LOP, together with a log reading.

3 Choose any point on the first LOP (point A in this case). Working from here, plot your DR from the course and distance run between the two LOPs.

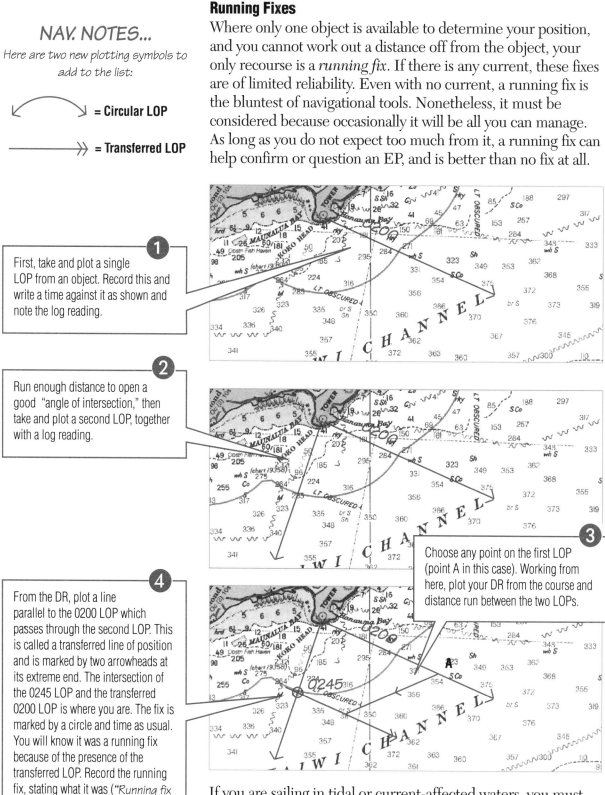

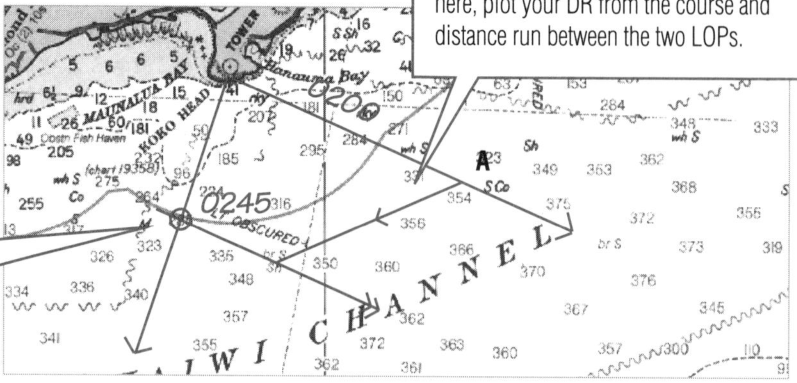

4 From the DR, plot a line parallel to the 0200 LOP which passes through the second LOP. This is called a transferred line of position and is marked by two arrowheads at its extreme end. The intersection of the 0245 LOP and the transferred 0200 LOP is where you are. The fix is marked by a circle and time as usual. You will know it was a running fix because of the presence of the transferred LOP. Record the running fix, stating what it was (*"Running fix on chart,"* etc.) and check the depth.

If you are sailing in tidal or current-affected waters, you must plot an EP from the point A in the illustration instead of a simple DR.

A running fix looks rather daunting on paper for the first time, but after you have executed a few in practice sessions, the technique will become second nature.

Defining your Position

It is sometimes useful to define a position in more absolute terms than merely as a circle and dot on a chart. You may, for example, wish to give the Coast Guard your position, or discuss plans with a "buddy boat" via radio or cellular phone. As well, electronic navigation devices generally refer you to position described by latitude and longitude. Here are the two principal ways of defining your position:

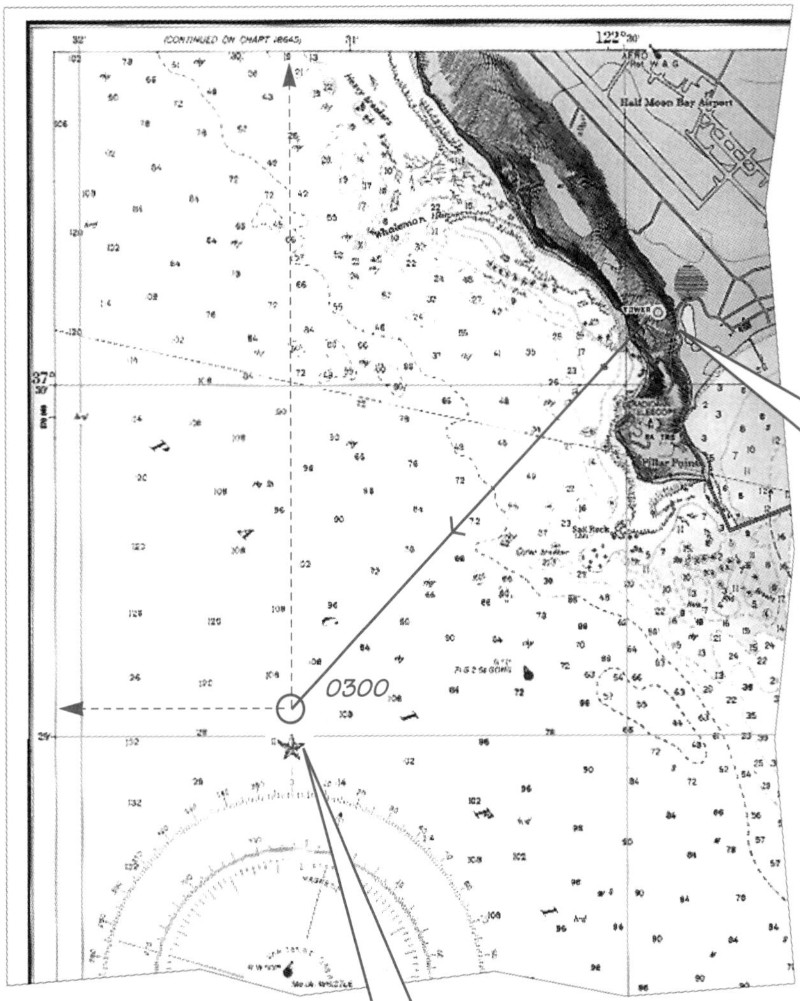

An alternate way of defining a position is as a range and bearing from a known point. Here, a tower near Pillar Point is used. The range is always in Nautical Miles, and the bearing is in degrees TRUE. If you are giving a position to rescuers and opt to use degrees magnetic, you MUST tell them, because all professional craft use True bearings.

This position is 37° 29'.1 North, 122° 31'.2 West. The easiest way to measure this in practice is to place a plotter or parallel ruler on one coordinate, e.g. the latitude, and read this off from the scale on the edge of the chart. With the ruler in place, use your dividers to determine the longitude.

CHAPTER 11 Electronic Navigation Aids

Radio aids to navigation have been available to yachtsmen since World War II, but only in the mid-1980s did they really come of age. A modern electronic fixing aid not only positions you with a high degree of accuracy, it can also give a course from one point to another destination. It can tell you your course and speed over the ground, and by comparing this with boat speed and course steered is able to deduce set and drift. Some will even plot your position on an electronic chart!

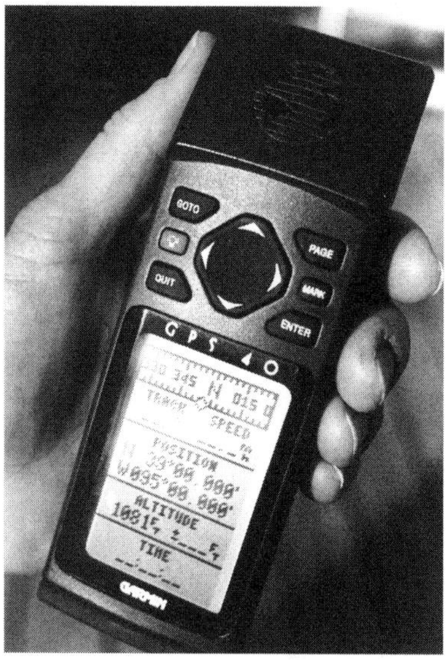

Ralph Naranjo photo

GPS receivers are extremely accurate, reliable and easy to use. Convenient hand-held models (above) have helped make GPS today's preferred fixing aid for pleasure craft.

A proper approach to these remarkable instruments is to integrate them into your overall navigational plan so that they are not critical to safety. That way, they can make life easier and help you navigate more accurately, but you will not be "up the creek" if they malfunction. Do not be misled by advertising or your own desires into believing that Loran or GPS can make a novice into a good navigator. It cannot, and anyone who insists otherwise is on course for a bad fall.

GPS

The *Global Positioning System* is a network of 21 satellites orbiting the earth in a "birdcage" pattern, with three more in orbit on standby. It is operated by the military, who may make periodic adjustments to the system (including interference with

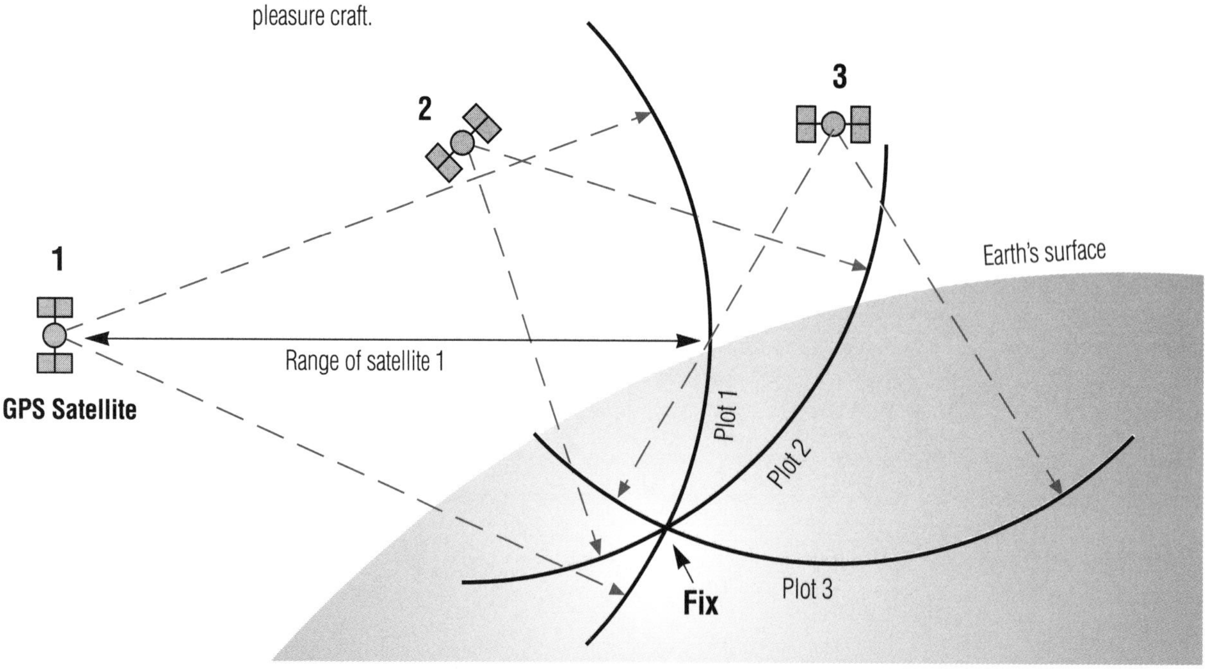

its accuracy), so theoretically the system should not be relied upon completely. In practice, GPS has proved extremely accurate, reliable and easy to use. It is now the preferred fixing aid for most yachts. However, it has been switched off in the past (in the interests of national security), and may be again. You should never wander up a blind alley where GPS is your only way out.

A GPS receiver "sees" a number of satellites at a time. Each transmits an accurately timed radio signal which the receiver picks up and processes, deducing its distance from the satellite whose exact position it knows. The result is a "sphere of position." The receiver finds the best three satellites for a good "intersection" of these spheres and then works out a fix.

All electronic fixes are plotted as a triangle with a dot inside. It is a good idea to write *"GPS"* (or *Loran*, *SatNav* or *Radar*) alongside them as you note the time. They are recorded in the same way as visual fixes.

Accuracy of GPS — If you are in the military and have access to the PPS (Precise Positioning Service), you can expect accuracy of around one meter. If you aren't, you must make do with the SPS (Standard Positioning Service), whose accuracy may vary from 15-50 meters. The authorities can also switch into Selective Availability Mode, which positions you within 45 to 225 meters — a nominal 100 meters.

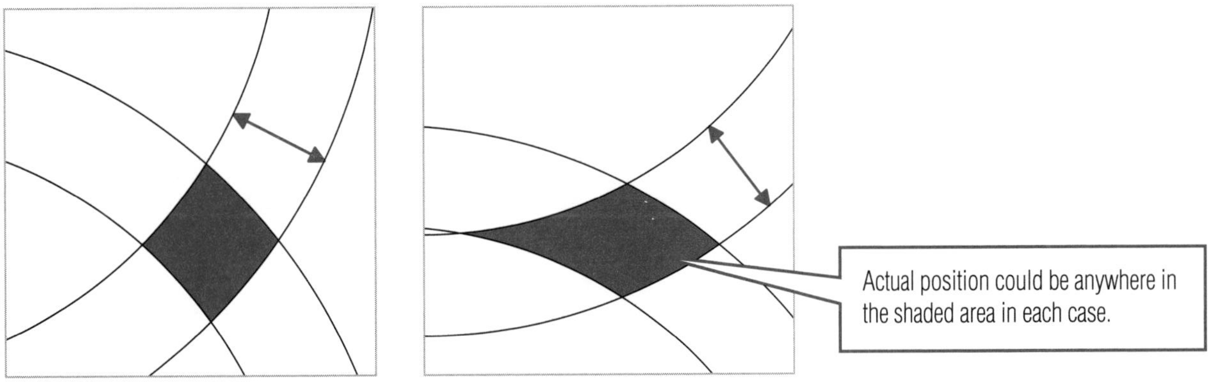

Actual position could be anywhere in the shaded area in each case.

GPS accuracy is sometimes affected by **Horizontal Dilution of Precision**, resulting from a poor "intersection" of the spherical position lines. When this happens, your receiver should either flash a warning or hold off computing fixes for a few seconds until the satellites have sorted themselves into a better geometry. The time gap will usually be of short duration.

Differential GPS — This is a method of upgrading an "average" GPS position to one of extreme accuracy. A receiver is fitted to your set which "locks in" to a nearby ground-based receiver, often a lighthouse. The position of the lighthouse is known precisely. If the receiver's GPS readout implies that it is somewhere else, even by a few meters, it works out the error in the GPS fix, then transmits the information to "differential" receivers in the vicinity so that the reading can be corrected. You won't know that this is going on. You just write the check for the equipment and plot your position to an extremely high degree of accuracy that the average yacht owner does not really need.

SatNav — the Transit Satellite System

Before GPS, the first satellite navigation system available to yachts was the so-called *transit satellite system*. This worked on the Doppler Effect of a radio signal from a satellite as it approached or retreated from the receiver. It was capable of reasonable accuracy, but suffered from the drawback of a small number of satellites giving a potential fix at quite long intervals - often as much as 90 minutes. This was adequate for ocean navigators, where the receiver, if linked to the yacht's electronic log and a compass, could work up a DR position based upon its latest fix. But such an instrument was dangerous if used in areas of strong current, as the DR computer took no account of the movement of the water. The system is no longer maintained.

Loran C

Loran (Long Range Aid to Navigation) C is a ground-based radio fixing system consisting of master stations, each of which has up to three secondary stations in its "chain." The onboard receiver computes LOPs of a hyperbolic shape by deciphering the phase difference between signals from the master station and its secondaries. Each set of signals generates an individual LOP, and there are usually three available from which the set can work out a fix. Compared with the European Decca equivalent system, Loran C is more powerful and covers a longer range. As a result there are a small number of chains operating on the continent itself.

Loran Station Chain

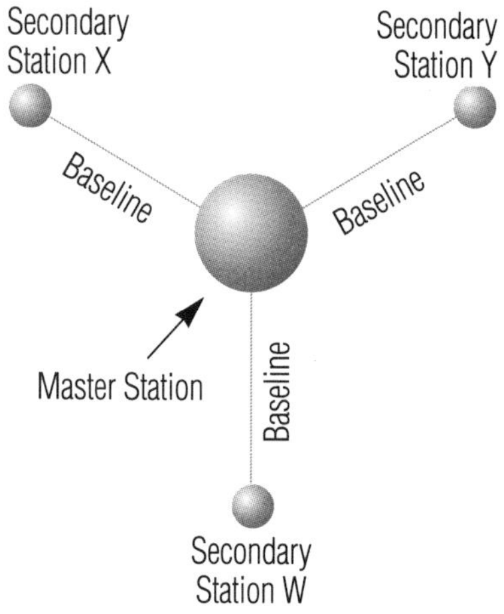

Loran C covers all of continental North America as well as many other sea areas of interest to American navigators. It is not, however, available in much of the Caribbean zone.

Accuracy of Loran C — If you are within range of direct "groundwave" coverage, Loran C generally assures you of a position fixed to within ¼ mile accuracy — often as good as 50 meters or less. Groundwave accuracy is good throughout the coastal regions of the US, including the Great Lakes. As you move out into the west coasts of Mexico and the Central Caribbean you come into "Skywave" country. In these areas you are receiving Loran C waves as they bounce off the ionosphere. Accuracy is degraded, sometimes to within two miles, with worst cases at 10+ miles. In the Eastern Caribbean, Loran is simply out of range, so you will have to depend on satellite navigation.

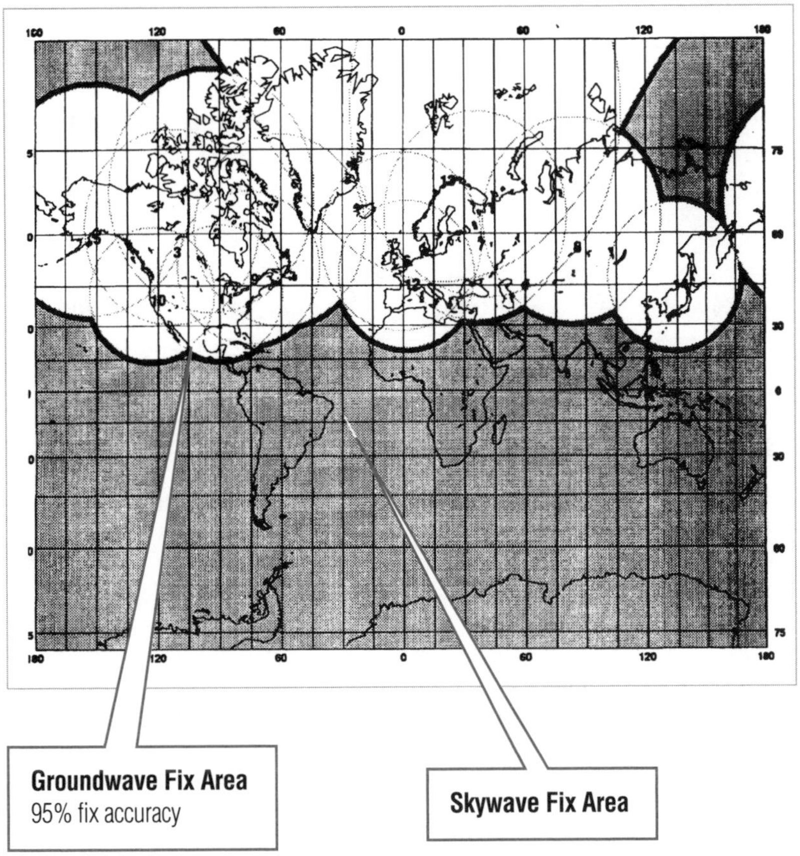

Fixed Errors — Like all radio waves, Loran C signals are bent by certain terrestrial features such as mountains and even powerlines. Such anomalies are tabulated in special books, but these are published only for use with commercial receivers. These sets give a value for the LOP from each secondary station which is then plotted by the navigator using a special "Loran C overlay" lattice chart. This refinement is normally denied the small craft user.

The line joining a master and a secondary station is called a *baseline*. In a few widely separated areas, errors may occur in the vicinity of a *baseline extension*, which is the area where the baseline from master to secondary extends beyond the secondary. *Loran C Reliability Diagrams* are available for each chain. These are designated as charts, numbers 5592 and 5606.

Variable Errors — Atmospheric electrical disturbances spell trouble for Loran C. A lightning storm can upset your readings or even temporarily knock reception out altogether.

Sometimes a station transmits dubious signals. When this happens, it may also send out a *blink signal* which, when picked up by a quality receiver, will activate a warning on the readout. When blink is occurring, positions must be treated with caution.

Repeatable Accuracy — If the position of a specific item (e.g., your local harbor entrance) as given by Loran C is different from that on the chart, do not worry. Because of fixed errors in the system, your Loran's position will be the same every time and so will anybody else's. This result is known as *repeatable accuracy*.

This feature makes Loran C particularly good if you are calling for assistance, but you must tell the rescue vessel that you are giving them a *Loran C derived position*. Whether or not it is strictly accurate is not important. If the rescue vessel plugs it in as a waypoint, its own Loran C will take it straight to your location. It is also useful if you need to return to an absolutely precise point.

Watch out for atmospheric electrical disturbances when using Loran C. A lightning storm can upset your readings or knock reception out for awhile.

Electronic Navigation Computers

All modern electronic navigation instruments, be they GPS or Loran C based, offer a number of functions in addition to fixing. Here are the most usual ones:

Course and Speed Made Good — These functions tell you the yacht's current course and speed over the ground. They are called *Course Made Good* (CMG) or *Course Over Ground* (COG), and *Speed Made Good* (SMG) or Speed Over Ground (SOG). Due to the effects of wind and current, however, CMG may vary from the course your compass says you are steering and SMG could vary from the speed or distance indicated by your log.

The log and compass can only measure boat speed and course steered through the water. The difference between SMG and CMG calculated by the navigation instrument and the readout of your on board log or compass is the result of current and/or leeway. By comparing SMG and CMG with your boat speed and course steered, you have a way to quantify set and drift.

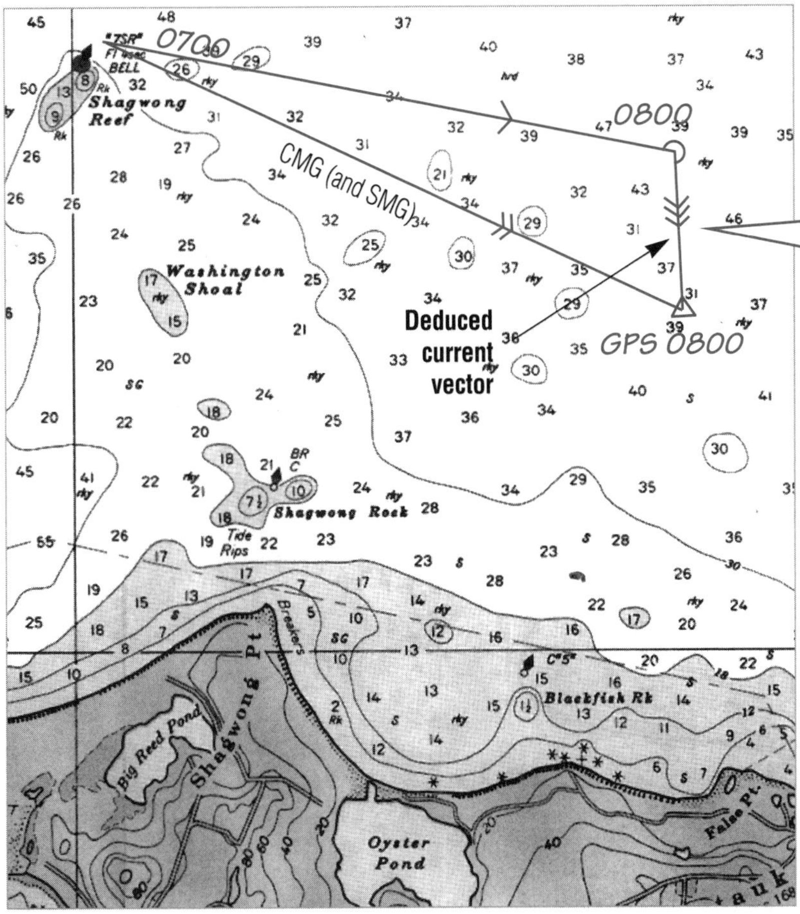

The best way to figure set and drift electronically is to draw a quick vector diagram. The course steered and boat speed form the "course line," while the CMG and SMG indicate where the boat has actually been — the track. Join the ends of the two lines, then give the resulting line three arrowheads because it is none other than the current vector — the set and drift (assuming your DR was corrected for leeway, if any). Measure its length and direction if you need to know the answer in numbers.

The end of a pre-planned route showing waypoints 7 and 8. The yacht's position at 0730 is defined as 4.2M from waypoint 7 on the opposite (reciprocal) of 185°. To find a reciprocal bearing, add 180°. If the angle is greater than 360°, subtract 180° instead.

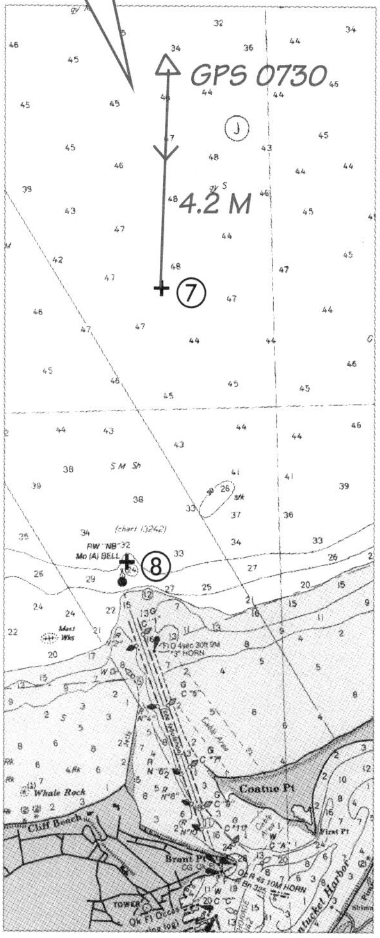

Waypoints — Navigators making full use of electronic navigation computers often work their way from one position to another down a pre-planned safe track. Such predetermined positions are called *waypoints*, and a large number can be entered into the average receiver in the form of "routes" or "sail plans." Range and bearing to the next waypoint are generally available on demand, which is useful in two respects: 1) If you stray from your direct line, you can see at a glance what track must be sailed to bring you back on course, and 2) you can plot your position simply by drawing a bearing from the waypoint and measuring the range with dividers. This is often easier than plotting a Lat/Long position, and in any case should be done where convenient as a cross check. It's all too easy to make a mistake while plotting an electronic fix. Many receivers can read out in True or Magnetic, depending on how they are set up, so READ THE INSTRUCTIONS!

Cross-track Error — This tells you if you are on the direct track from one waypoint to the next. It also indicates how far off course you are and which way to steer to come back into line.

Such information can be valuable if navigating in poor visibility, but it is a mistake to become obsessed with staying precisely on track when it is not important. *Cross-track error* is useful for double checking the effect of current once you have already calculated your offset (see Chapter 7). If you opt to use the cross-track error function from the outset, rather than plot a vector diagram, you will make a lot of course changes before you arrive on the correct course to steer. Do not use cross-track error on a long cross-tide passage which will involve a major change of current direction. Chapter 7 will show you why not.

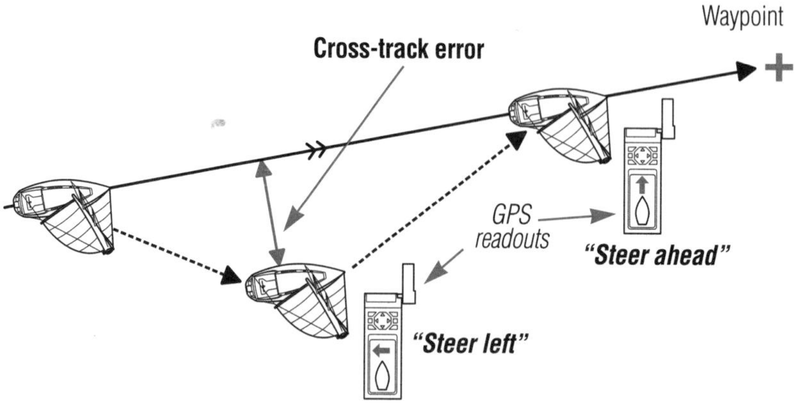

GPS-induced Collisions — If you are sticking slavishly to a well-traveled track defined by the super-accuracy of GPS, remember that some other boat may be doing the same, headed in the opposite direction. In fog this can be particularly dangerous, so keep a sharp lookout at all times.

Overboard — Most receivers have a "man overboard button" which instantly records your position as a waypoint, over-riding all else and giving a constant readout of course and distance to the victim. This can be a lifesaver, but don't forget that the fix recorded is a ground position that does not account for current. If you are in 2 knots of current, the victim will have drifted 400 yards away from the recorded position within the first 6 minutes after the incident.

Human Error — Voltage may fail from time to time and the GPS network could occasionally be switched off, but by far the biggest source of electronic grief comes from operator error. It is too easy to press the wrong button when punching in a waypoint, or to plot a wrong digit when transferring a fix to the chart. Take your time, be careful, be precise.

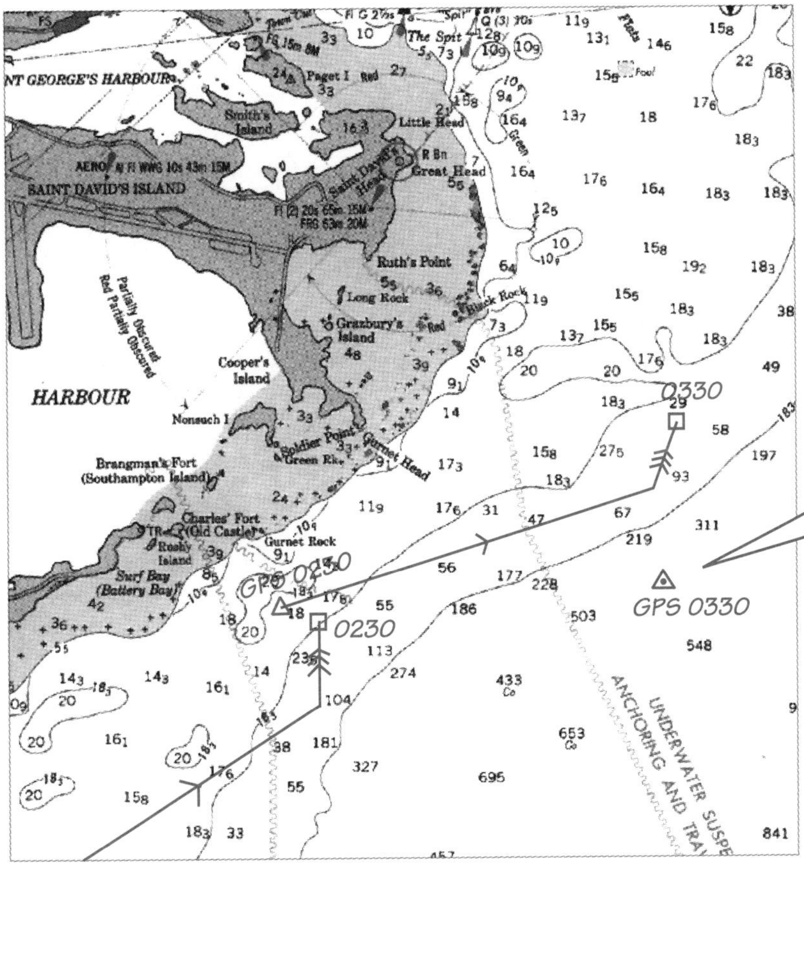

Always double check everything you do. Keep up your conventional plot as well as your electronics, and jump on any anomaly immediately. This illustration shows a GPS fix that has suddenly gone adrift from the plot. The immediate action is to take a sounding. If that does not show the fix is wrong, you will have to delve deeper...*but dig you must until the truth is revealed!* Failure to check the plot would have given this navigator a false sense of security.

Radar

Radar is the most interactive aid to navigation, and requires a higher level of skill from its operator than any of the electronic fixing aids because its readout is not in unambiguous digits. The rewards are great for those who persevere, however, because not only does radar supply a view of navigational features which cannot otherwise be seen, it is also a primary tool for collision avoidance. In this book, however, we are interested in the instrument purely for its navigational capabilities.

Radar Bearings as LOPs — Most radar sets have a rotating *Electronic Bearing Line* (EBL) running from the center of the screen to its edge. Once you have identified a charted object on the screen you can rotate the EBL until it touches the object and read off its bearing from a digital display. If your radar gives a "head-up" picture (i.e., the top of the screen represents your ship's head), the EBL bearing will be relative to your heading. NOTE: *Some radar displays may be oriented with North at the top, meaning that bearings taken by the EBL are compass bearings.*

Radar bearings are not ideal as LOPs for two reasons. 1) They are often not very accurate because the boat yaws around its heading, and 2) if the scanner sends out a beam width of, say 4°, a bearing can be inaccurate by this amount. If you are sighting on a small target, placing the EBL in its center will help.

If you fix your position using radar bearings only, the same rules apply as to LOPs from bearings taken by eye. Look for a good intersection, use closer rather than distant objects, etc.

Radar Ranges as LOPs — As we have seen, bearings taken from a radar set may be far from accurate. The capacity of radar to determine distance off, or range, with its *Variable Range*

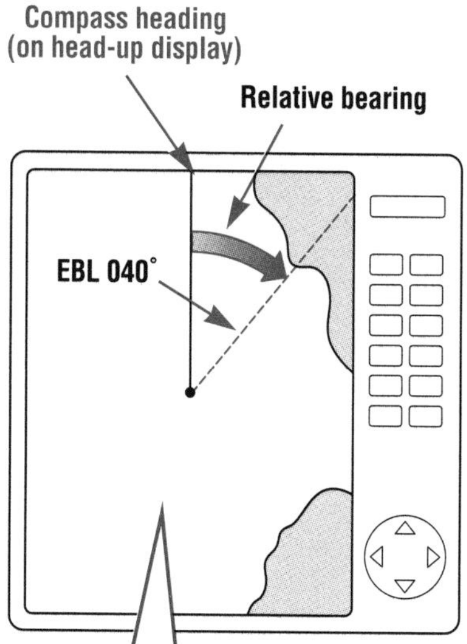

Converting an EBL reading to a compass bearing on a "head-up" set is a straightforward matter of adding the relative bearing to the compass heading.

Compass heading	**135°M**
Relative bearing	**+ 040°**
Compass bearing	**175°M**

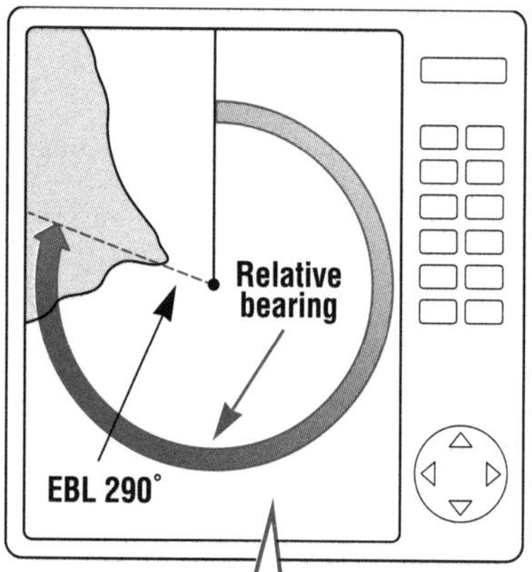

Where the sum of relative bearing and heading is greater than 360°, subtract 360° from the result to come to a usable answer.

Compass heading	**135°M**
Relative heading	**+290°**
Compass bearing	**425°M**
	– 360°
	= 065°M

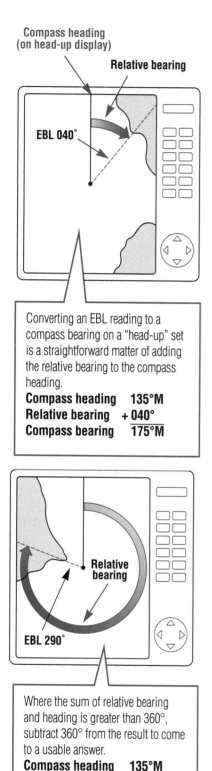

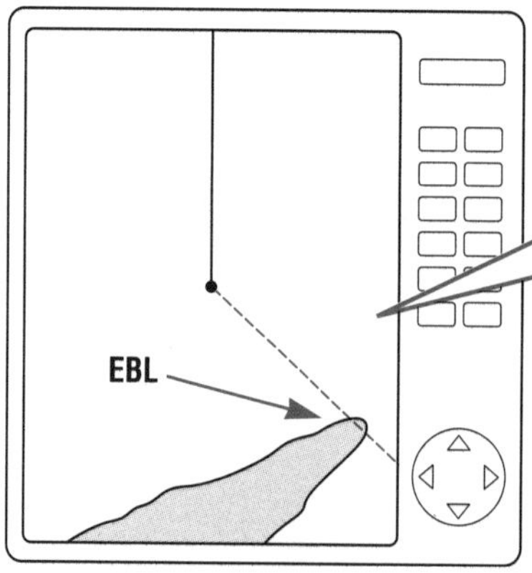

Headlands and similar large targets viewed obliquely generally appear closer than they are by about a half-beam width. Therefore your bearing will be more accurate if you place the EBL at about a half beam width to landward of the edge of the image. Some sets can adjust their beam width to minimize this effect, but if not, a total width of 4° is a fair assumption unless otherwise stated.

Marker (VRM) is far more refined. It is generally good to within ±1% of the range scale in use. A single distance from an object can be plotted on the chart as a circular LOP just as you would for the distance off a rising light. If you can produce three such LOPs, you have a three-point fix whose cocked hat can be evaluated as though it were a visual fix.

Mixed Radar Fixes — If only a single target is recognizable on the screen when you need a radar fix, you can produce one of sorts by using the EBL and the VRM on the same object.

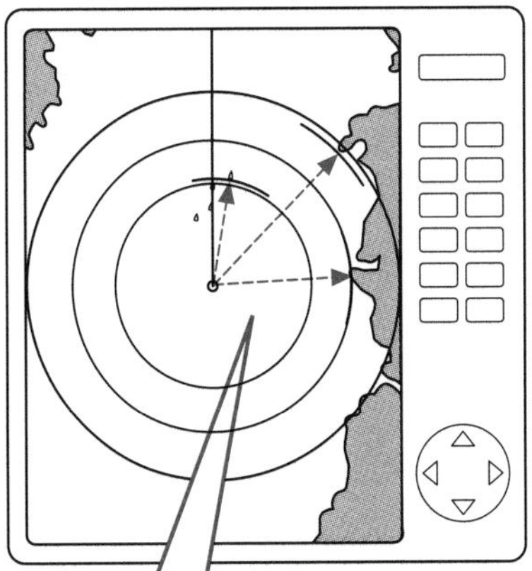

Choose three prominent radar targets and note their distance from you by using the VRM. Read the log. **1**

2 Now set your drawing compass to the distance of each object in turn and, centering the point of the instrument on the first object, scribe the arc of a circle in your approximate vicinity on the chart. Now do the same with the other two objects. The intersection of the three LOPs is your fix. Mark it with a triangle as an electronically derived position, and don't forget to record it in your ship's log..

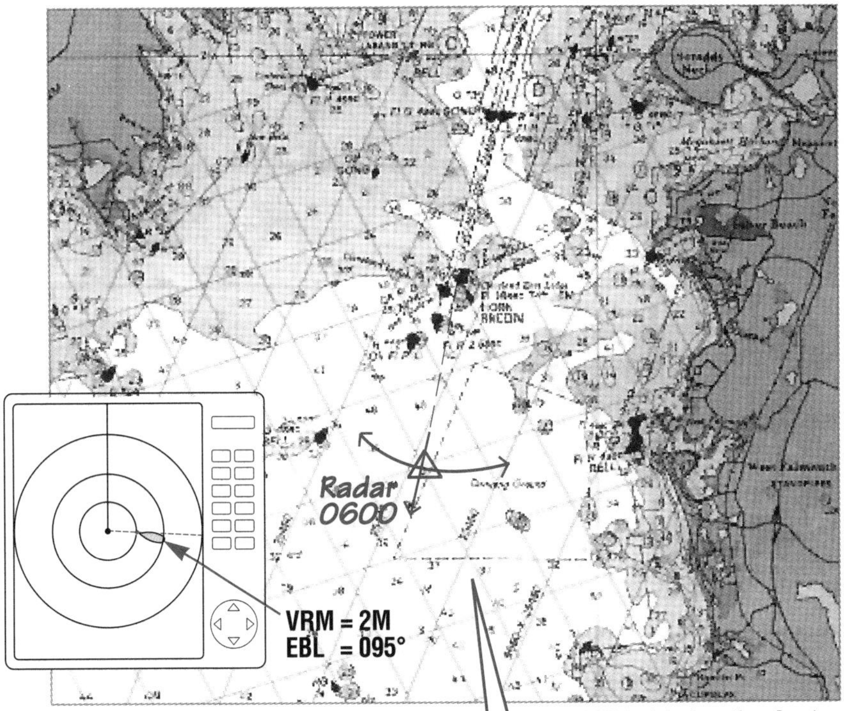

NOTE: Actual chart does not have a Racon buoy.

Ship's Head 275°T
Relative Bearing + 095°
= 370°
− 360°
= 010°T

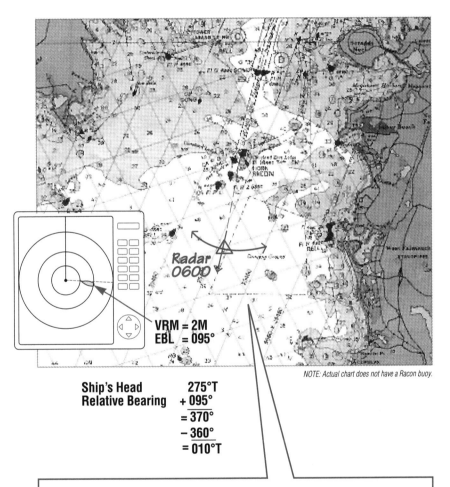

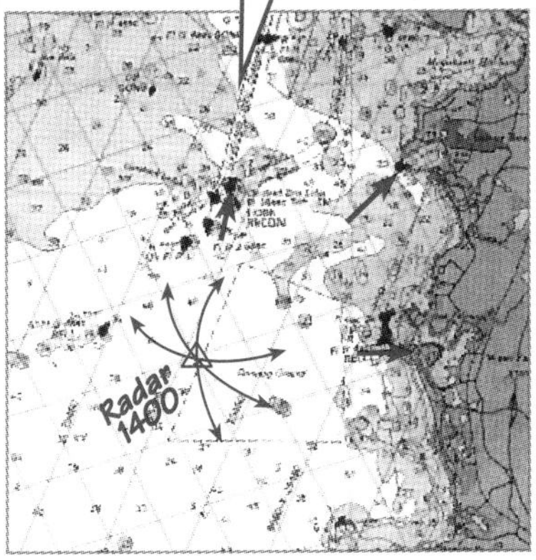

Plot the distance off as one LOP on the chart, then the bearing as a second. Within the limits of the radar bearing, you have a fix. Record it and look for a better one. Here, the object in view to the radar is a Racon, which is a radar reactive target marked as such on the chart, in this case a light float. Its compass bearing is 010°T (ship's head + relative bearing) and its range is 2 miles. The time is 0600. This racon identifies itself by a flash running radially out behind the object "blip" on the screen. Others may show a Morse code. Had it been possible, a better fix might have been a visual bearing crossed with a radar range. A third LOP from some other source, even the fathometer, would have further refined the position.

CHAPTER 12 Inshore Pilotage

As you move into waters crowded with buoys, beacons and hidden dangers (worse still, hidden dangers and no buoys!) situations can develop too quickly for formal chartwork, so piloting techniques must change gear. This is especially true in heavy weather, or in areas where currents run hard or are unpredictable. Such conditions often occur at the beginning or end of a passage where you are sailing in unfamiliar harbors or anchorages, but it can also be necessary to execute a tight piece of piloting en route from one place to another.

Special procedures are required for this type of piloting. They must be simple to set up and follow because in tight quarters, hesitation often spells failure. Fortunately, the art of inshore pilotage revolves around a few basic principles from which you can develop your own approach

The Safe Course or Heading

The essence of inshore pilotage consists of knowing that you are sailing from one safe position to another on a straight line that is clear of danger. You do not necessarily have to know where you are on that line. Inshore pilotage is not about position fixing, it is about single, safe LOPs. If the LOP doesn't hit the rocks and you are on the LOP, you are safely on your way. .

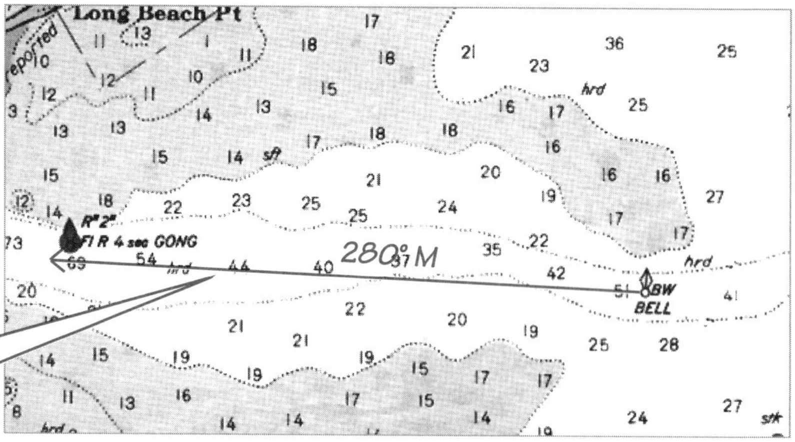

In this example, an LOP runs from a fairway buoy into a narrow channel on a "safe course," between shoals on both sides of the channel.

The Clearing Bearing (or Danger Bearing)

Another form of LOP used in inshore piloting is the *clearing bearing*, or *danger bearing (see example at top of next page)*. Clearing bearings are not a course line or the heading of the boat, but instead define the edge of a safe area. When using a

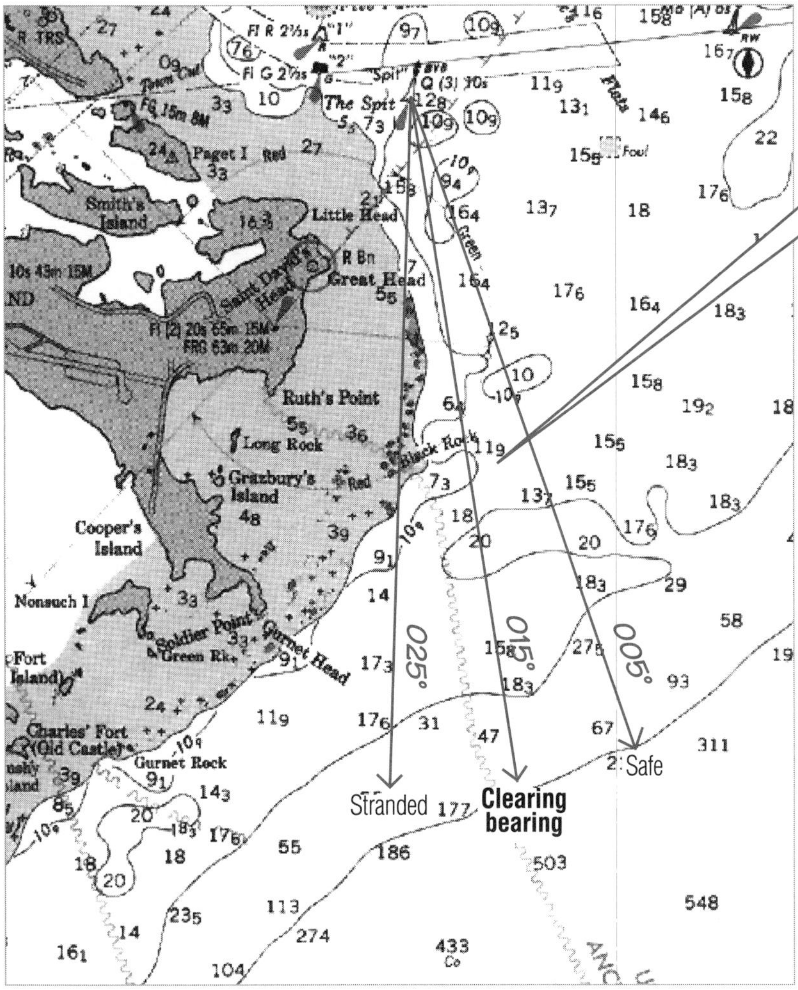

This **clearing bearing** is a LOP running along the edge of a reef. If the buoy bears "greater" than 015M you are on the rocks. So long as it bears "less," you can sail on. *Remember, clearing bearings are not a course line or the boat's heading.*

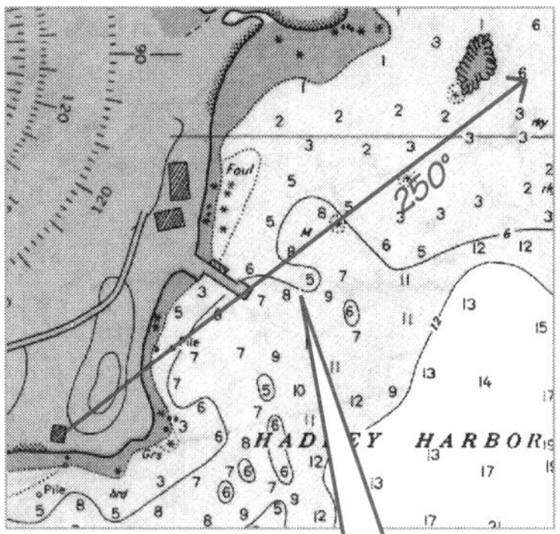

This clearing bearing is a range (*transit*) defined by a house onshore and a dock. It is a far better way of sighting a clearing line than a compass bearing because it is so well defined.

This **official range** defined by flashing onshore light towers, is a charted LOP defining a safe course through a narrow channel.

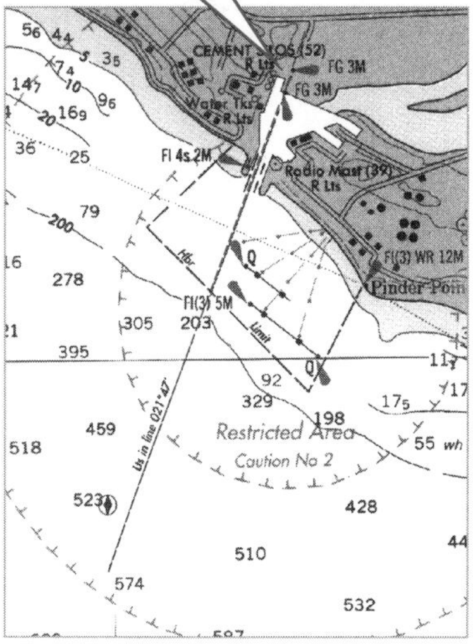

Ways of Defining an LOP

Sources of Lines of Position for inshore pilotage are the same as in any other form of navigation (*see Chapter 10: "Knowing Where Your Are" for examples*). Your key to success is knowing beyond a doubt that you are sailing on the line you want until you are ready to move on to the next.

Unlike offshore navigation, the compass is the pilot's last resort in close quarters. The best technique is always a *range*. GPS might tell you your position within 20 - 30 meters, but a good range can tell you which side of your cockpit you are sitting on. A typical "official" range would be the flashing onshore lights defining a safe course for entering the narrow Freeport Harbor Channel (shown right). If there is no official range, try to find one of your own by reading the chart creatively, just as you would when seeking an LOP.

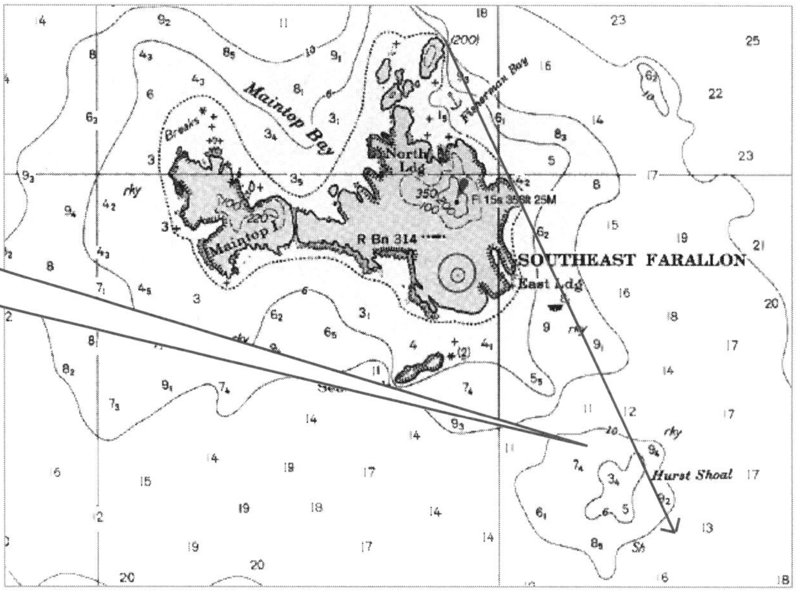

An Unofficial Range as a Clearing Bearing

Here, Southeast Farallon Island and a smaller island behind establish a clearing bearing for Hurst Shoal. If the small island behind Southeast Farallon disappears you are in trouble. As long as it is aligned ("on") with Southeast Farallon, or "open" to it, you are safe.

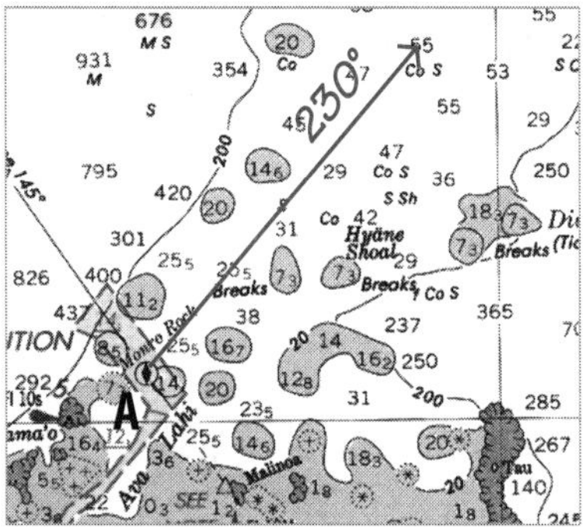

The Back Bearing

In the absence of a decent range, it is often necessary to operate from a back bearing to a single object. In the example to the left, it will be necessary to note a back bearing to buoy A, which will be a reciprocal of your desired track. The reciprocal of 050° is 230° (50 + 180 = 230). Sight your hand bearing compass on the buoy (see below) as the helmsman steers away on 050°. Keep the compass on 230°. If you see the buoy drift left or right, adjust your course to bring yourself back onto line. If, on the other hand, you stay sighted on the buoy and note the change in bearing—then work out which way to turn by doing a sum—you can get into trouble by adding when you should have subtracted, or vice versa.

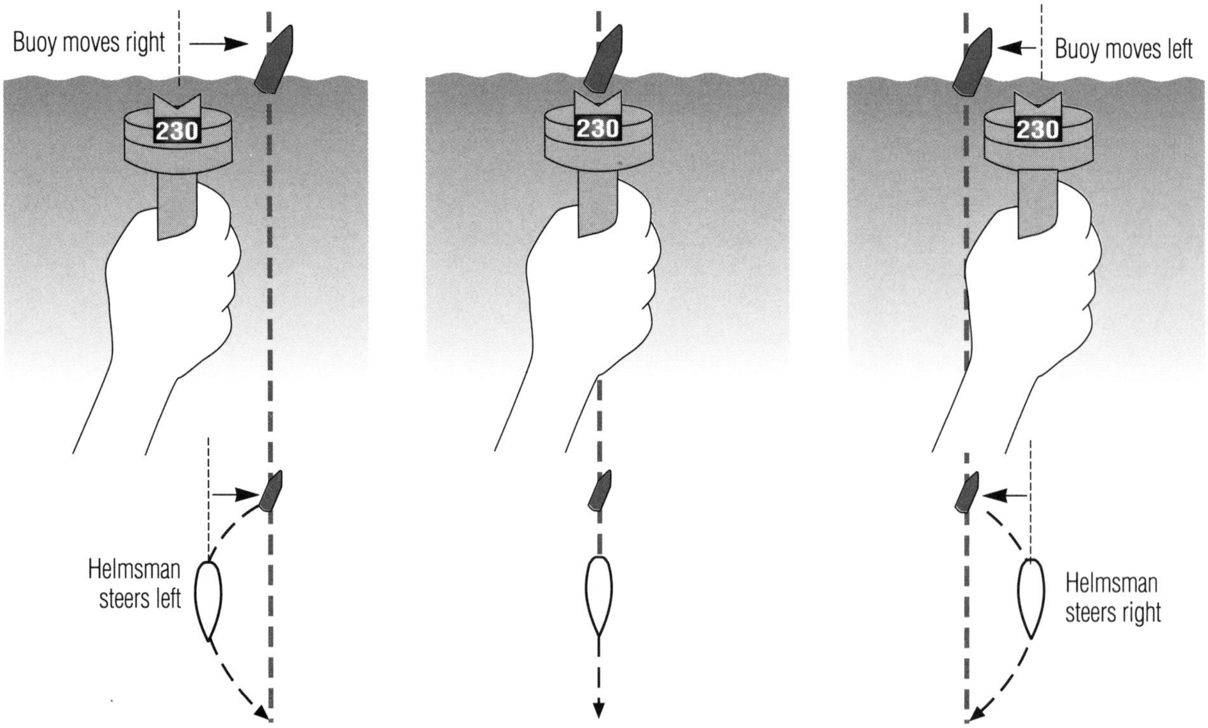

Steering a Bearing

Occasionally you will have no choice but to steer towards an objective on a safe bearing. The concept is similar to a back bearing, but in reverse. Using your steering compass, keep the boat on the required heading and note whether the objective stays dead ahead. If it drifts to one side, and the boat is still heading 090°, you can see instantly which way to "twitch" the boat in order to get back on line. Steer left in this case, then adjust your course slightly and see if the objective now stays in line. Making further adjustments as required until you are satisfied.

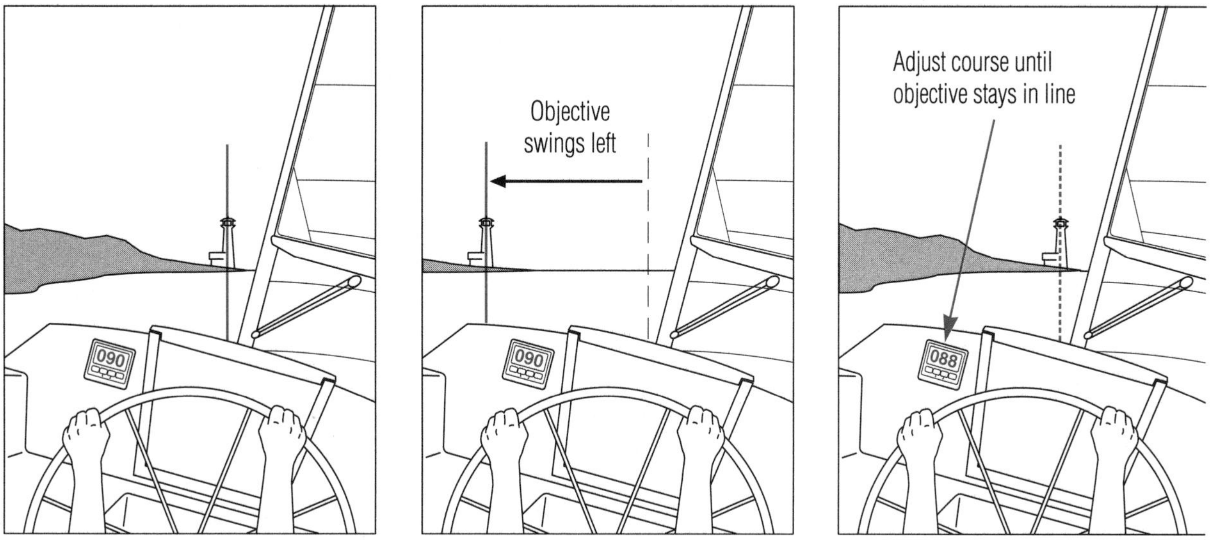

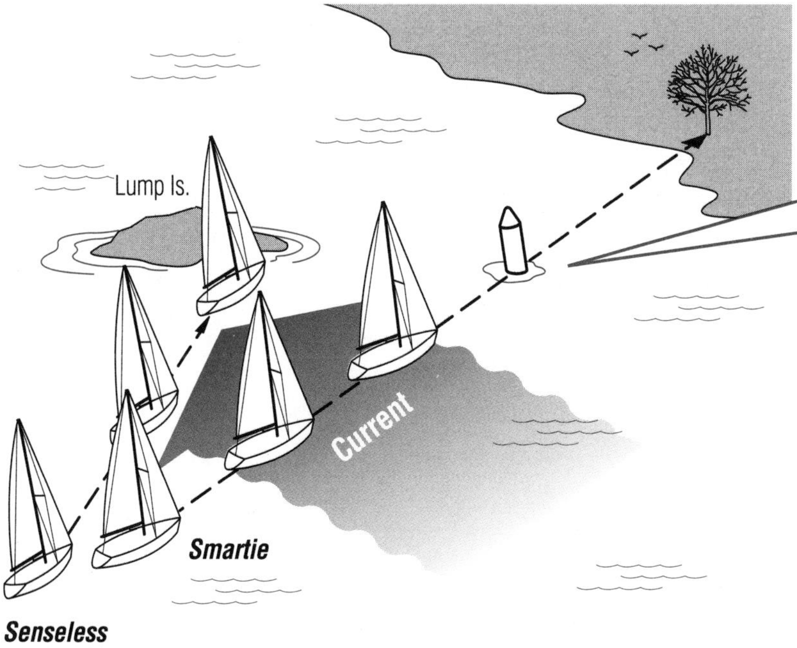

Proving a Compass Bearing

In a cross current it is not good enough just to round a buoy and steer towards the next one. If you do, the current may drift you into danger. Here, *Senseless* has done just that, and ended up aground on Lump Island. *Smartie* has "ranged" the next buoy on a tree (it could have been any stationary object). This means that *Smartie* is certain to keep exactly on the safe bearing.

Use of the Fathometer

The fathometer is a vital pilotage tool. Beating up this unmarked channel, for example, it is giving two safe lines to tack on. As the boat rounds the bend onto a reach, it can easily keep to the 18 foot line to maintain a safe heading on the correct side of the channel.

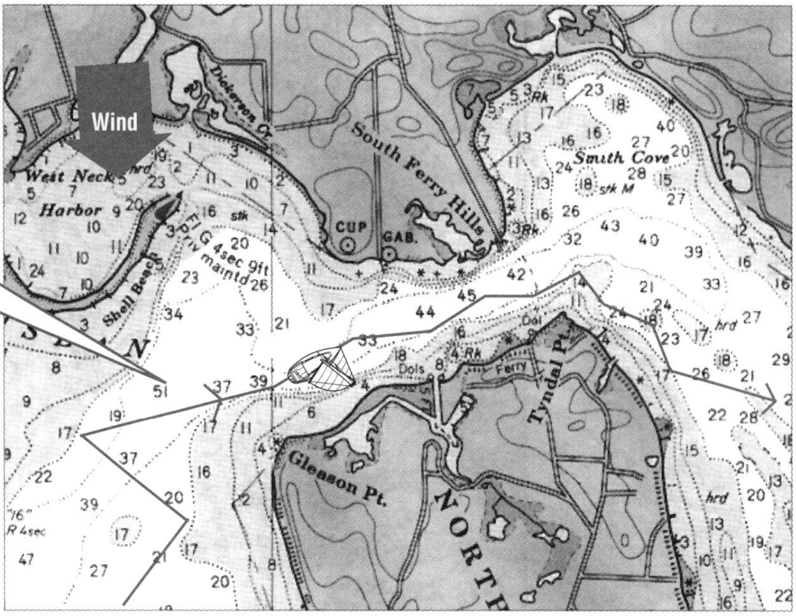

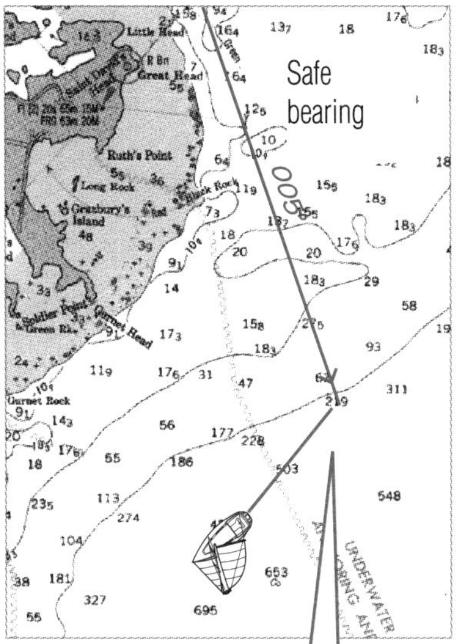

The fathometer can also tell you where you are entering or leaving danger when running a safe bearing or staying outside a clearing line.

Objects Abeam

You can judge when an object is exactly abeam to a surprising degree of accuracy by sighting along your mainsheet traveler, or down the aft edge of a hatch, or anything else running square across the yacht. To determine the bearing of an object abeam, add 90° to the ship's heading if it is on the starboard side, or subtract 90° if it is to port. This can be handy as a source of instant LOP if you need one, but it is more often used as a cutting-off point for a clearing or safe line.

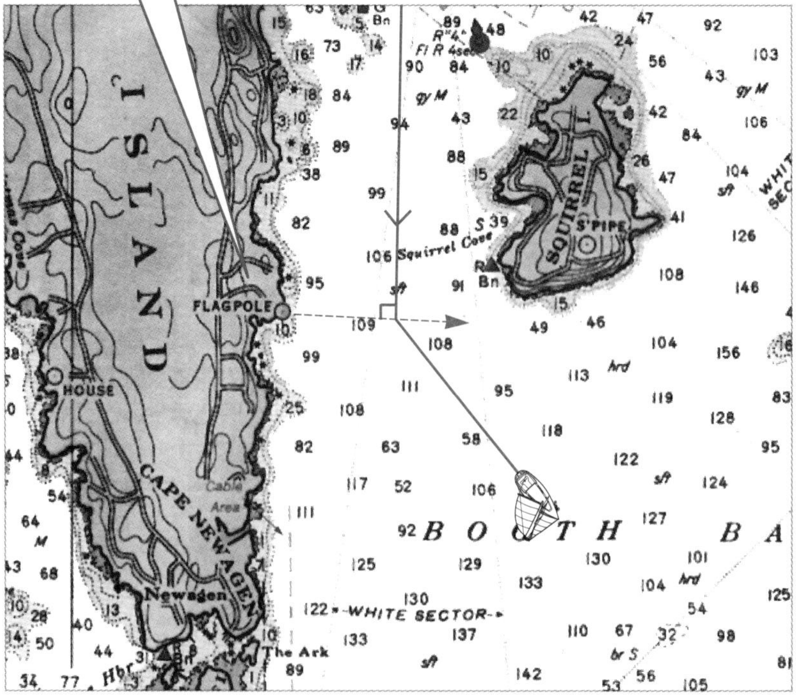

Tropical Eyeball Pilotage

In areas of extremely clear water it is often possible to see shoals and to make successful judgments regarding the nature and depth of the bottom. Always check your chart and sailing directions before "eyeballing" a critical reef and do not expect great things early and late in the day. A high sun is an important part of the package, as is a good pair of Polaroid sunglasses.

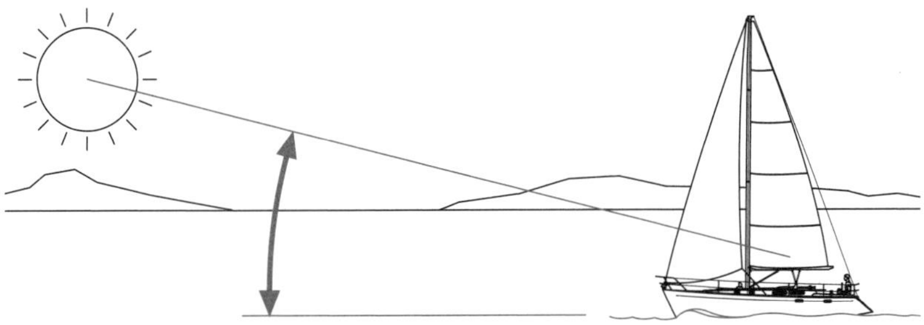

Eyeballing shoal waters with the sun low and ahead can be dangerous (see above) because visibility beneath the water's surface is diminished. A high sun, especially abeam (see right), is ideal for sighting the colors of the seabed. Position a crew member at the bow or standing on the cabin top to call the colors from a higher viewpoint.

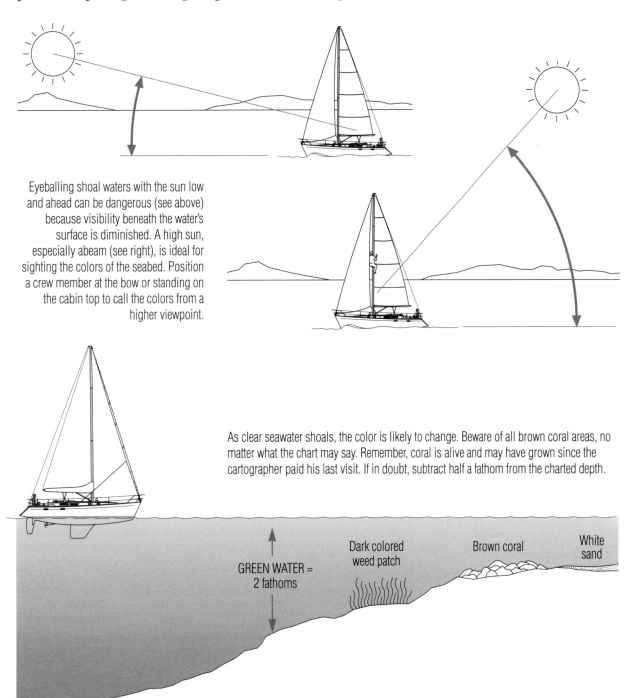

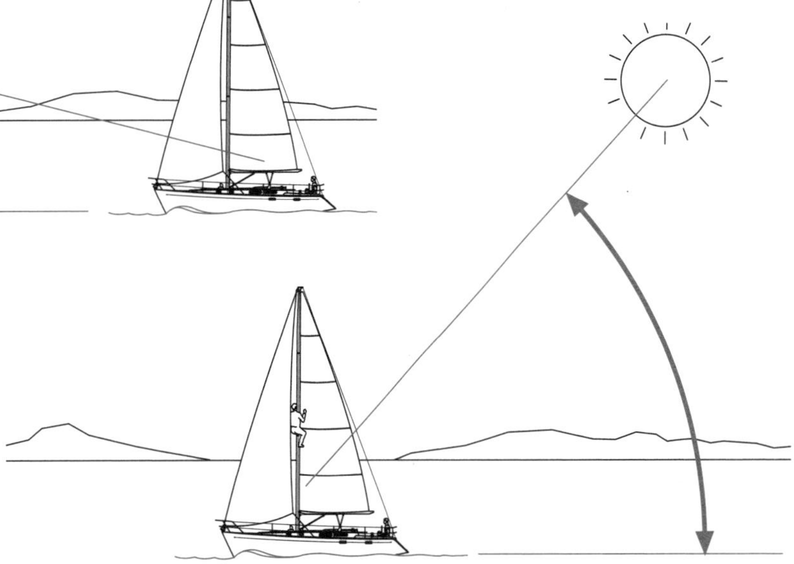

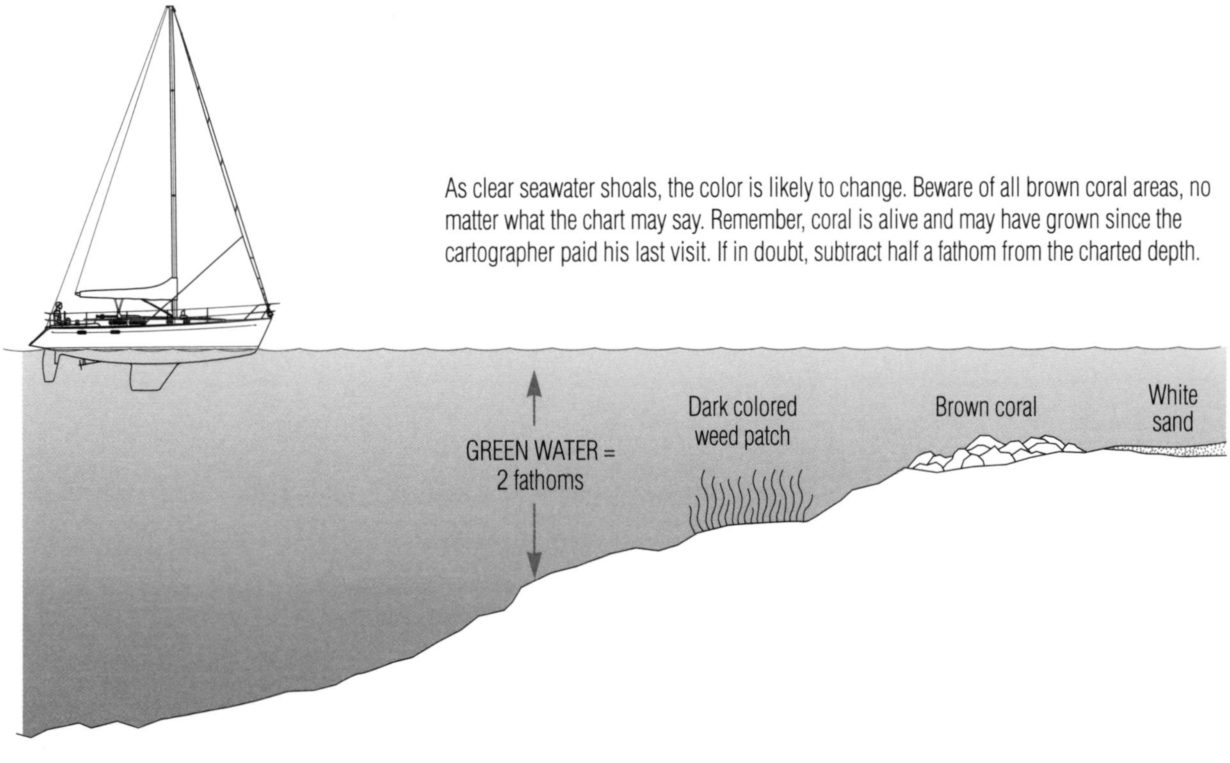

As clear seawater shoals, the color is likely to change. Beware of all brown coral areas, no matter what the chart may say. Remember, coral is alive and may have grown since the cartographer paid his last visit. If in doubt, subtract half a fathom from the charted depth.

Practical Inshore Pilotage

Some cruising guides present you a pilotage plan for entering an awkward anchorage. If so, it will offer a series of bearings, courses and sometimes distances to take you from one safe spot to the next. If there is no such guide available you must organize your own plan. How you do this is entirely up to you as long as

1 From *Middle Rock* buoy, steer 274M, keeping on track by heading for the house on *Bear Island* (A) or watching a *back bearing* on the buoy. Keep checking your back bearing and whatever happens don't let it drop below 084 M as this *clearing bearing* (or *danger bearing*) "clears" the submerged rocks north of *Eagle Island*.

2 As soon as C "1" off *Fling Island* bears 241 M, swing to port toward it. You have only your compass to help you here, so it will be useful to line up the buoy on this bearing with a handy reference like an uncharted house on the right-hand side (RHS) of *Oak Island* (B) or a rock that shows at low tide, then use this as an unofficial range. Once past the buoy, a back range can be formed between the can and the RHS of *Butter Island* (C).

3 Just after you have come abeam of *Sloop Island* and it lines up with the high hill on the left-hand side (LHS) of *Great Spruce Head Island* (D), you can proceed down this *natural range* steering 172 M .

4 If you don't like the looks of the rocks between *Dagger Island* and *Oak Hill*, you can swing to port when the RHS of *Bald Island* bears 111 M and steer a couple of degrees off the point at 113 M.

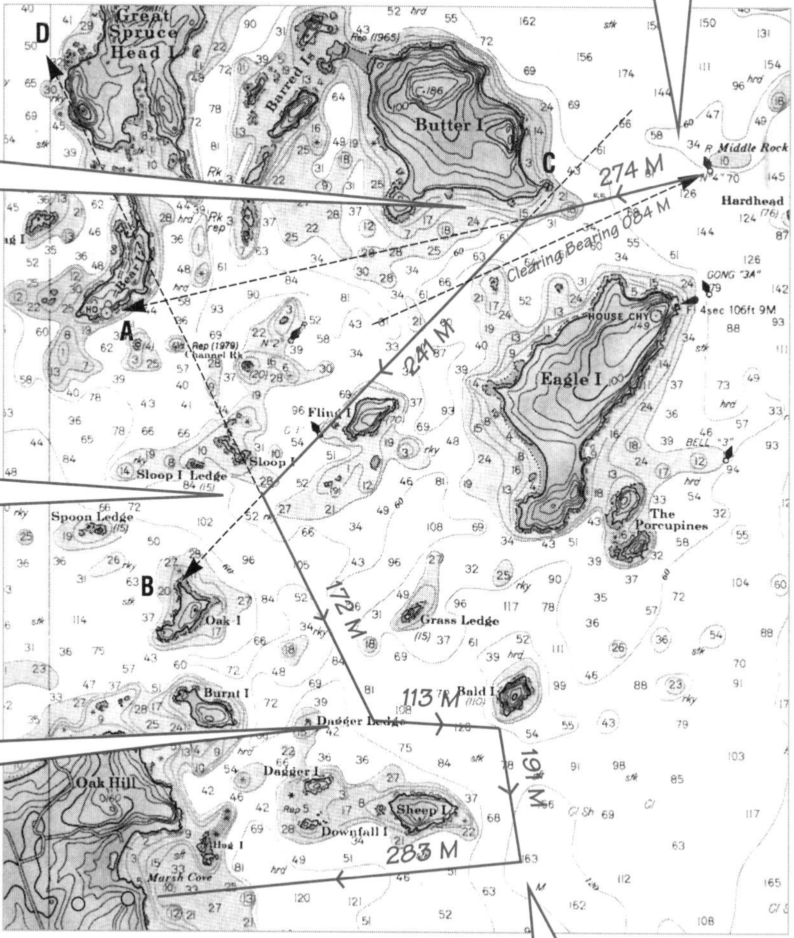

NOTE: Actual chart does not have range markers at Marsh Cove.

5 At *Bald Island*, come to starboard and steer 191 M which can be checked with a back bearing of 011 M off the LHS of Bald Island. Leave *Sheep Island* to starboard and come onto the "official" *Marsh Cove Range* on 283 M. As the depth shoals toward the 3-fathom contour, you can swing to starboard and anchor.

you do not end up having to plot courses or measure distances during the action. Some people make up a plan in a notebook. Others write notes on the chart, which they then keep in hand. Either is OK, so long as you have a system which you can understand and which tells you all you need to know. Make up a plan early, particularly if you will be entering a strange harbor after dark.

Night Entries

Sound inshore pilotage techniques can stand you in good stead for entering strange harbors at night. However, the closer in you come, the greater your reliance upon your eyes to see more than just the lit buoys, so it is vital not to become overconfident. In daylight, you can see if there is an uncharted or unexpected object in your path, such as a large fish marker with its attendant nets. In the dark you cannot. Without background light, all you can make out are lit navigation aids and sometimes the reflective marks on the unlit ones.

Many harbors have enough background shore lighting to be able to move around safely once "inside" the limits of the navigation aids. Some do not. Use your common sense and if in doubt, stay at sea or go somewhere else.

In spite of this advice, never be over timid. It would be a shame to lose your yacht on a lee shore by refusing to enter a safe harbor which had been available before the weather got too bad, just because night had fallen.

You know the techniques. Use them with the greatest care. Spot all the buoys, move at a safe speed, and don't be afraid to stop, either by anchoring or heaving-to, if things start to "smell" wrong. Take a good look around you. Fix your position if you can, reassess your situation, wait for your heart to slow down, then press on in safety.

CHAPTER 13 Navigational Strategy

Navigational strategy is not really necessary for the motorboat operator or the big ship skipper. For them, it is enough to steer the shortest geographical route between two points, the *rhumb line*. For sailboat navigators, there is a little more to it than that.

Yachts travel faster and more comfortably on certain points of sail. Tactics and strategy for the sailor are all about maximizing time spent on a reach, or at least laying the course, and minimizing the dead runs and those seemingly endless periods pounding into oncoming waves with the destination planted on the weather bow. This is achieved by applying good sense and anticipation so as to accommodate changing weather.

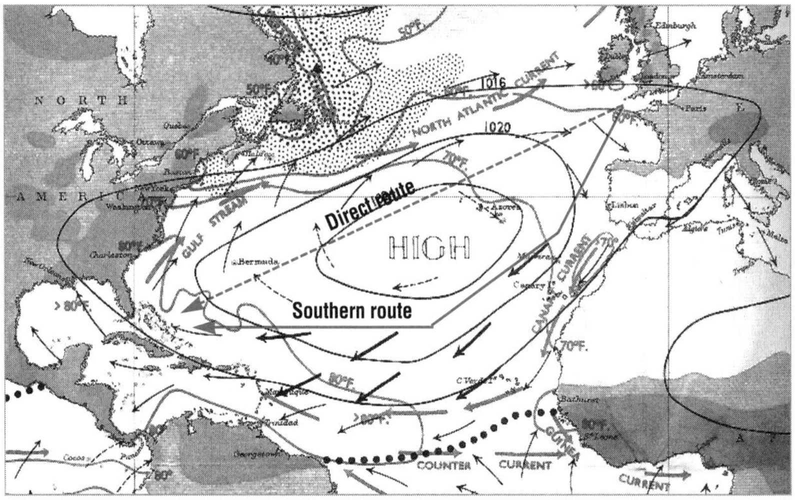

A classic example of navigational strategy comes from a Trans Atlantic passage. No properly briefed sailor would set out to sail a direct route from the English Channel to the Caribbean (see right), battling first with westerlies, then with calms. Instead, as Trans Atlantic sailors know, he or she sails a more Southern route, down to the trade winds, then enjoys a fair wind all the way across at the small price of an extra thousand miles or so. This is smart strategy at its simplest.

The Upcurrent Arrival

Another example is projecting a landfall following a crosscurrent passage of some hours' duration. A Gulf Stream crossing to the Islands would be typical.

If you are sailing out of sight of land and do not wish to rely exclusively on electronic fixing, you are well advised to plan your course upstream of your destination. After you have a *landfall fix* it will be easy enough to alter course and cruise in with the stream behind you. If you steer straight for your destination and end up downcurrent you'll surely kick yourself, particularly if you have to sail hard on the wind for the final stretch. A good general rule is to *keep upcurrent (or upwind when there is negligible current) if there is the slightest chance you may not lay your destination.*

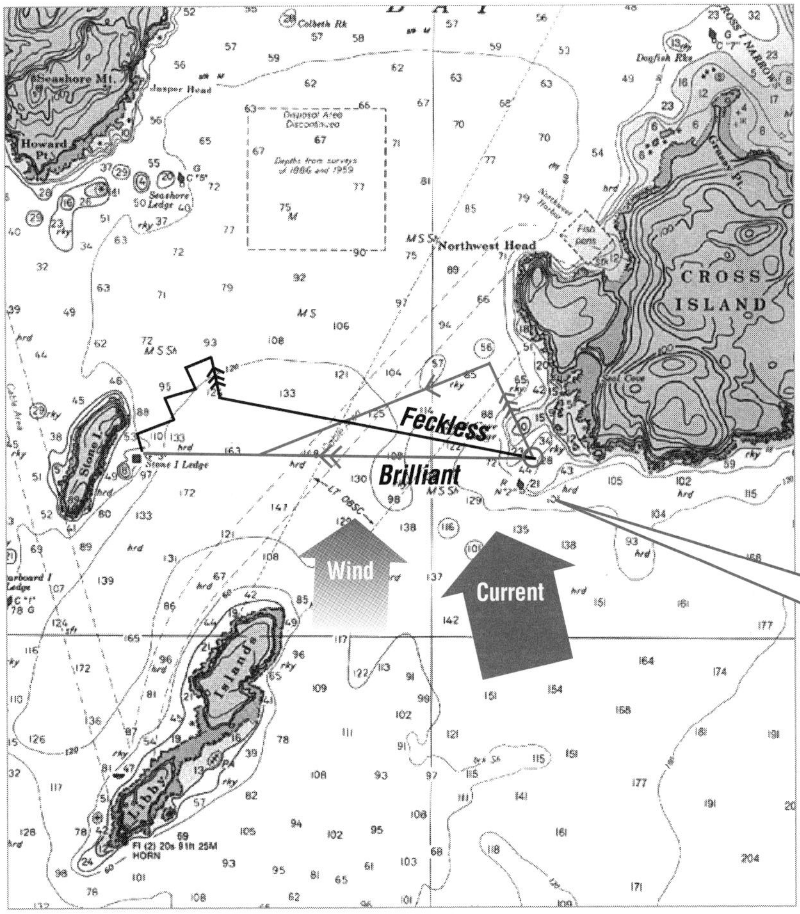

Upcurrent Arrival

Here, *Feckless* has foolishly steered directly for Stone Island, been set downwind by the current, and now must beat back against it. *Brilliant* shaped her course upstream from the outset and will be relaxing at anchor long before *Feckless* is even in sight.

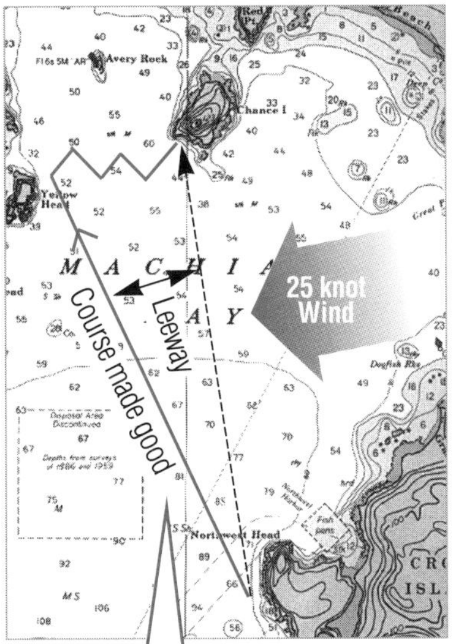

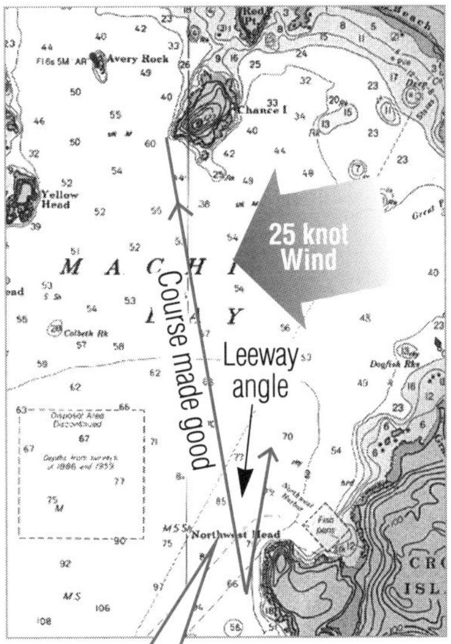

Upwind Arrival

Feckless has done it again, being murdered by her own leeway.

By applying leeway to her course beforehand, *Brillant* cruises in for an easy trip.

Upwind Tactics

When beating to windward you should spend as much time as possible on the tack that takes you closest to your destination, which is called the *favored tack*. If you are setting out with a longish beat ahead of you, always choose the favored tack to begin with. 1

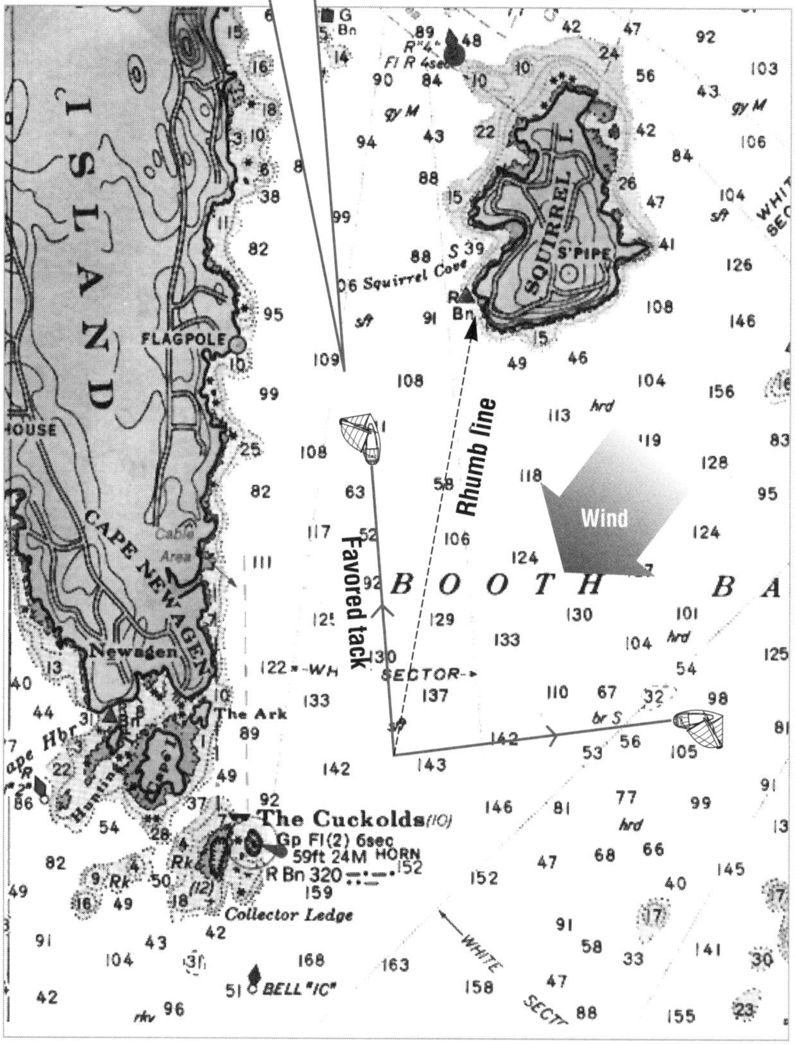

If your destination lies dead to windward, you will soon have moved out to one side of the rhumb line. Depending on the distance to be made good, you will find in due course that your present tack is taking you away from your destination, rather than towards it. After you have gone about, your new tack is the winner until you are once again squarely downwind of your destination.

If you're on the favored tack and get a favorable wind shift, you may not have to tack at all. 2

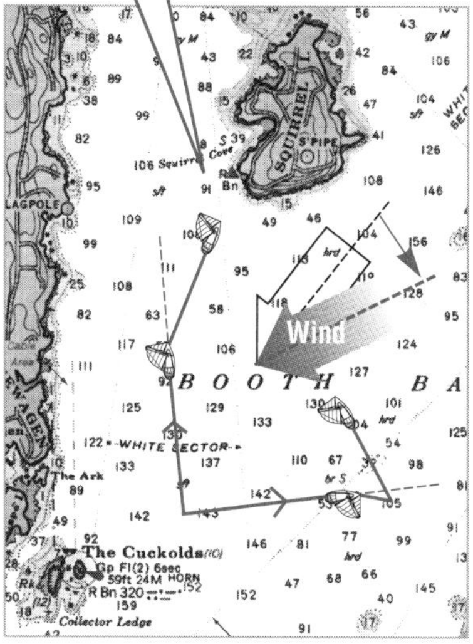

Even if you are headed, you can always tack and nothing will have been lost. In either case if you had put in a short losing tack first "to get your weather gauge in the bag," it would have been distance wasted. 3

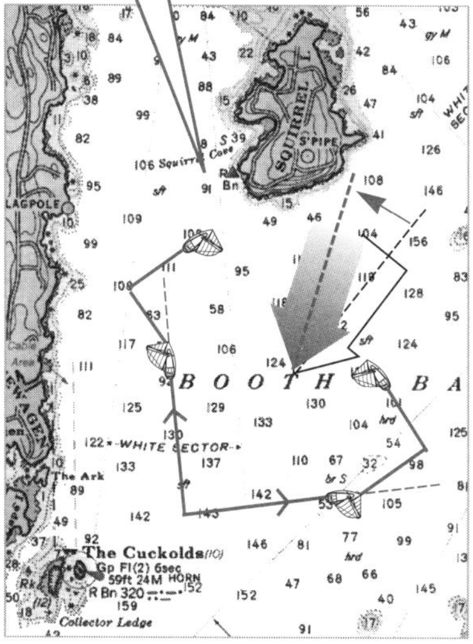

If the wind were constant you could in theory make a success of any beat with a single, perfectly timed tack but alas, the breeze shifts, sometimes predictably, often not. For this reason —

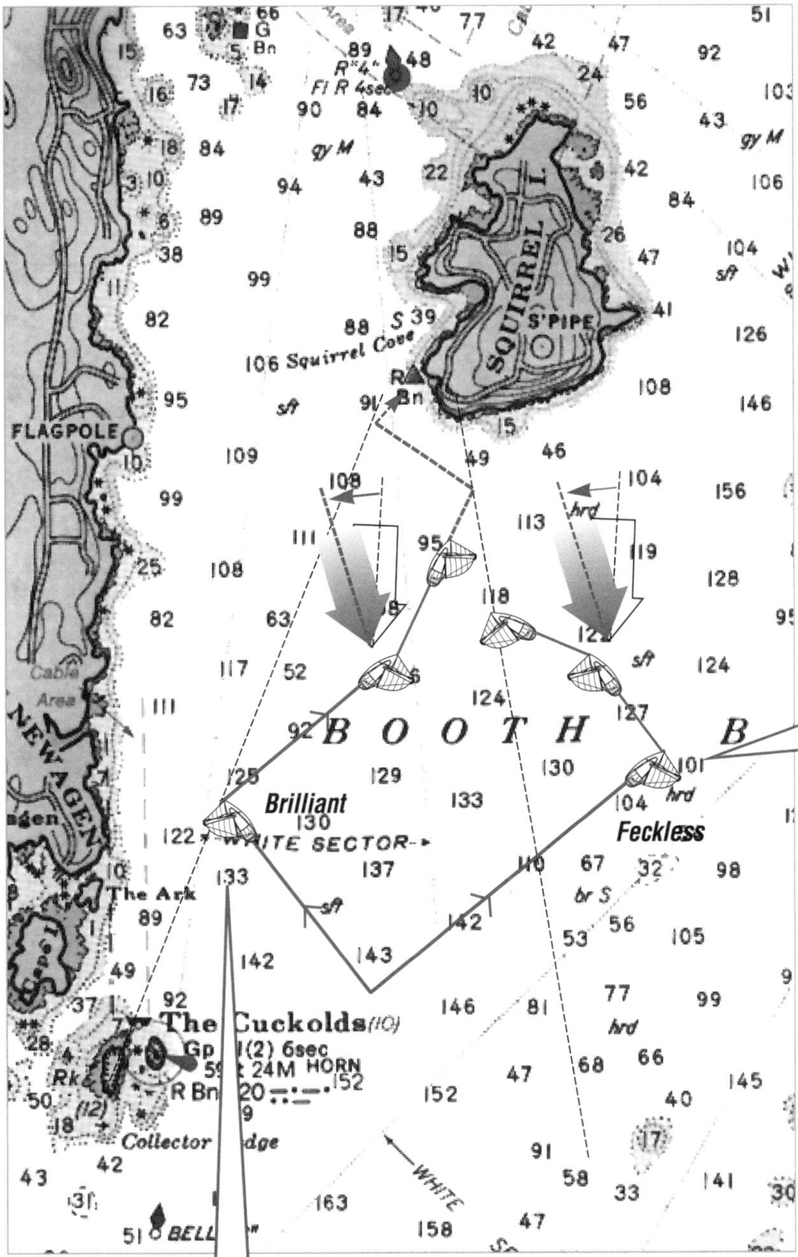

Avoid laylines

It is a bad idea to to windward on a 20-mile tack, waiting until you can lay your destination (called the *layline*) before coming about. If the wind were to shift after you tacked you would either no longer able to lay the destination or would be sailing free on the new tack, having sailed farther to windward than needed.

When to Work the Middle

The smart way is to tack up an ever-tightening cone toward your destination. This will ensure that you are best placed to maximize any advantageous shift, and minimize a bad one. If you do not stray far from the rhumb line, you can tack on any headers and gain immediate access to the better slant. If freed on an existing tack, you can stay on it for as long as it remains advantageous.

Go to the Favored Side for Anticipated Wind Shifts

On windward passages where you are expecting a wind shift, make sure that when it arrives you are on the favored side of the rhumb line. The new wind will carry *Brilliant* straight home, while *Feckless* finds herself far "down the tube" dead to leeward.

Downwind Tactics

Avoid a dead run if at all possible. Running in open water produces an uncomfortable rolling motion and (except in strong winds) is slow as well. It is often helpful to adapt a form of upwind tactics to downwind sailing in order to keep you on a fast, enjoyable broad reach.

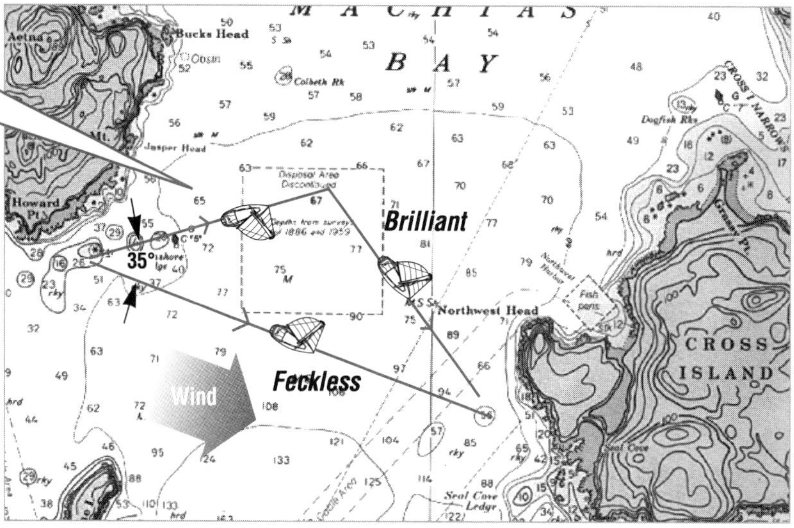

On a dead run of 60 miles, *Brilliant* alters course 35° to one side of the rhumb line, giving her an excellent reach. *Brilliant* can jibe as often or as infrequently as she chooses, but will add only about 14 miles to the overall distance. This may well be made up in increased boat speed, but even if it isn't, the whole trip will be much more enjoyable which, after all, is what we are out there for. Meanwhile, *Feckless* is sailing dead downwind "rolling her stick out" and probably upsetting a few stomachs along the way.

Light and Darkness

Always try to arrive at a destination in daylight if at all possible. If this is not feasible, use the light of the moon wherever you can. A good almanac will give the times of moonrise and moonset, but if in doubt, remember when the moon came up the previous night and add approximately 40 minutes. A night entry to an open anchorage will be far less stressful if you manage to arrive an hour or more after moonrise.

The Dawn Landfall

Sometimes, it is easier to make a landfall on a well lit but unfamiliar coast in the dark than in daylight. In daylight, one hill or headland can be mistaken for another, but there is nothing ambiguous about a lighthouse. Historically, most strandings have occurred because the navigators "knew exactly" where their boats were until they hit the shoals they thought were somewhere else.

If you can time your landfall for dawn, you enjoy the best of all worlds: the darkness or twilight shows up the lighthouses and any offshore buoys, while the growing daylight guides you safely into a harbor whose identity is certain.

Chapter 14 Navigation in Poor Visibility

Officially, fog is defined as visibility of 1100 yards (1000 meters) or less. In other words, half a mile or so. In a yacht unassisted by electronics, this is not very much when approaching a landfall after a 50 mile passage, but when the atmosphere thickens up to 50 yards visibility or less, even "Electronic Man" will be feeling a certain tightening around the stomach. And so he should be, for the first message about operating in fog is that it is unwise to rely entirely upon any piece of equipment which could fail for reasons outside your immediate control. In the context of fog, electronics must be looked upon as a valuable aid to conventional navigation. We shall therefore first consider the question of how to cope with fog from the position of a yacht with no Loran, no GPS and no radar.

Navigational Actions To Be Taken When Entering Fog

Your first action when you see fog developing is to fix your position and record it in your log. Even if you have electronics this may be the last chance you'll get to reassure yourself that the "black box" is functioning as it should. If you can't produce a fix, work up an EP as accurately as you can, then double check it against your electronics if you have any, in order to confirm both it and them.

Safety Checklist in Fog

- ☐ Double the lookout
- ☐ Hoist the radar reflector
- ☐ Listen hard for sound signals from navigational aids and other vessels
- ☐ Set a radar watch if appropriate
- ☐ Sound your fog signal (— • • under sail, — under power) every two minutes
- ☐ Everyone on board should wear PFDs in case of collision.

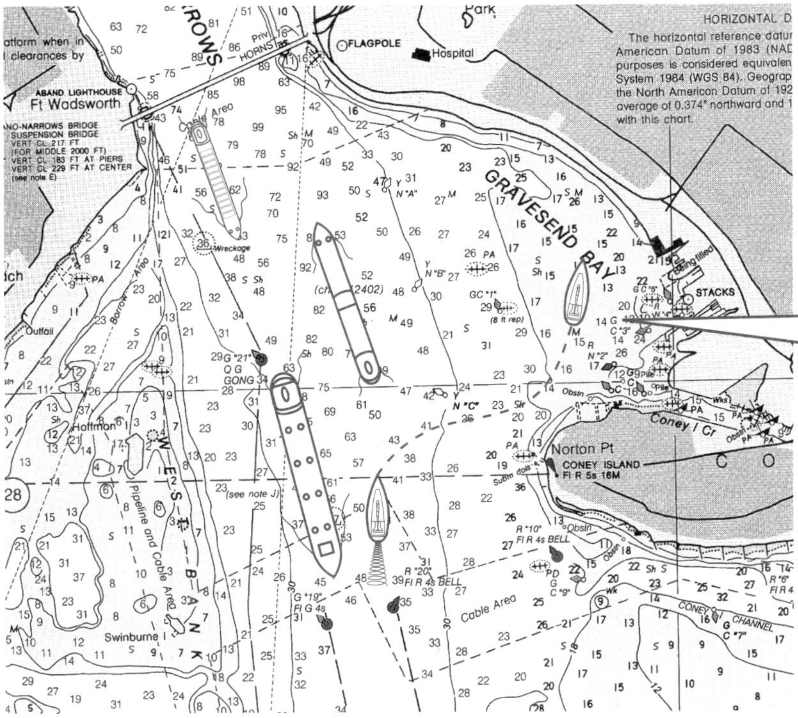

Fog Tactics

If you cannot make a safe harbor and there is heavy traffic around, you have two options: either sail far out to sea and heave-to or, preferably, make for shoal waters inshore and anchor in an area so shallow that a commercial ship could not hit you even if it tried.

Running the Contour

Should you be out at sea and lose sight of your landfall, do not fall into the trap attempted by *Optimist,* especially if there is a cross set. *Optimist* has attempted to make directly for the entrance, but has drifted to one side. *Optimist* will be faced with the question, "Which way must I turn to find the entrance?" It is a far better strategy to purposefully miss the entrance, but know which side you are on, like *Winner. Winner* can sail toward shore until she finds 24 feet, then turn to starboard and run the contour, secure in the certainty that she will come to the bell buoy at the harbor entrance.

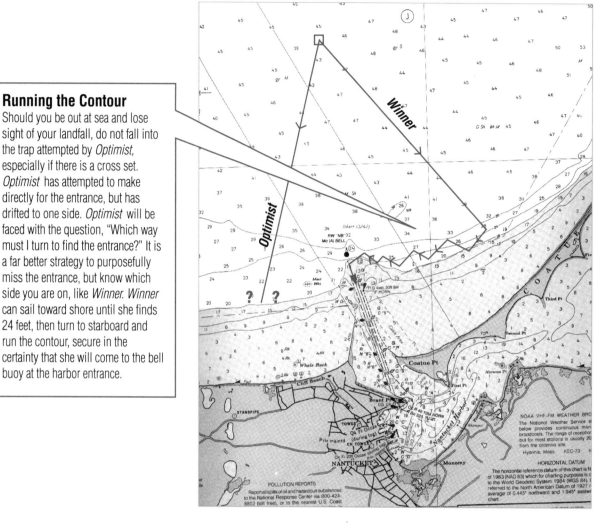

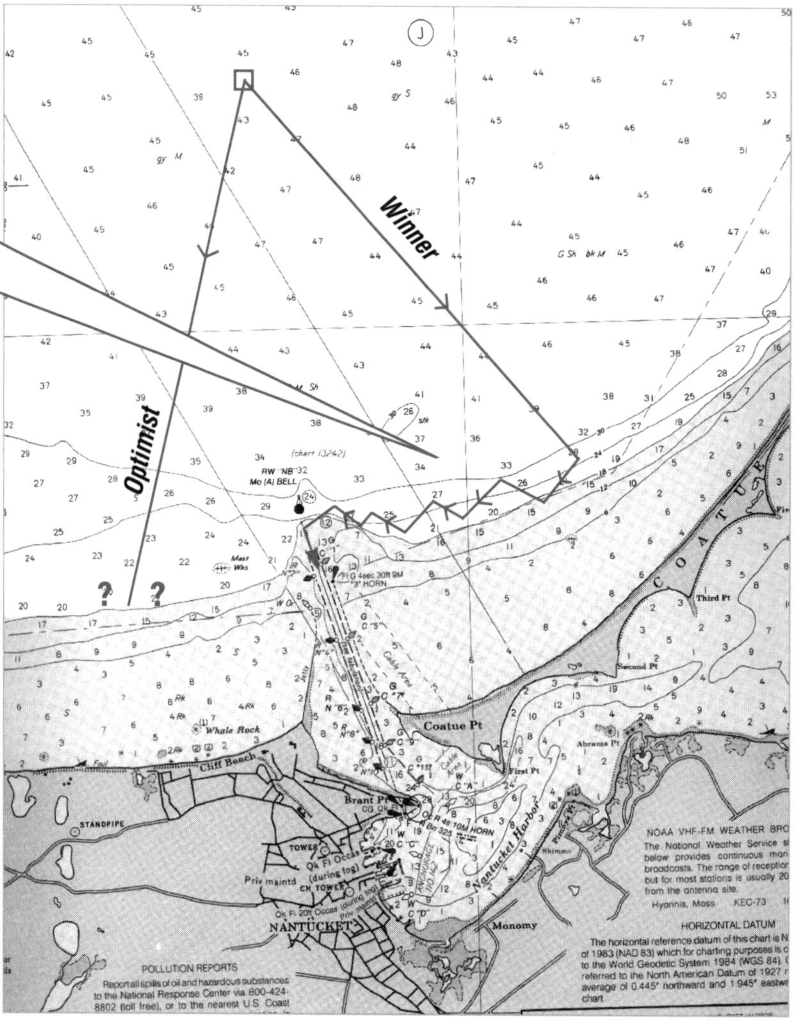

Buoy Hopping

This technique is also used in good visibility situations, such as entering an unfamiliar harbor. Running from one buoy to the next can keep you in touch with where you are, so long as you are reasonably sure you can find each buoy. Should you miss one, you will rapidly become disoriented because as soon as you have "run your distance" and the buoy has failed to show, you will have no idea where it is relative to your position.

Even if you are unsure of succeeding you can still try buoy hopping in safety, given two requirements: a contingency plan to bring into play if any of the buoys do not appear, and an assurance that failure to see a buoy would not immediately place you in danger. If you cannot come up with a safe course of action to implement if your plan should fail, you may be better off not trying the buoys, but using some other method.

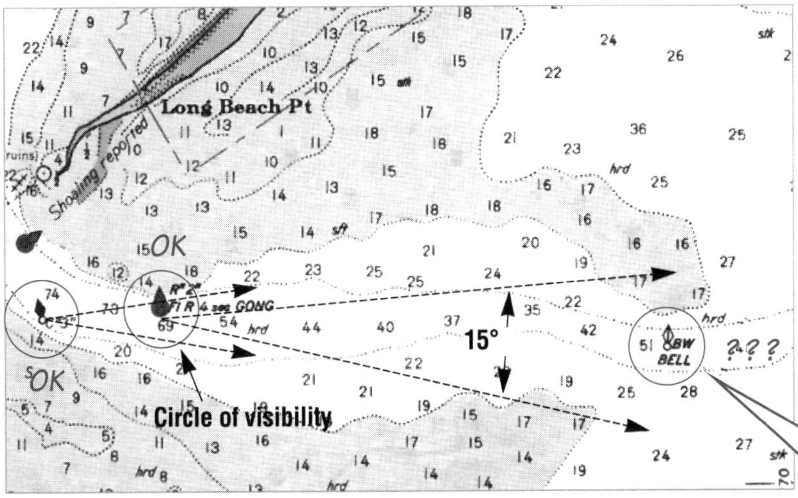

Steering for Visibility Circles

In the absence of current, most boats can be steered to within 7 or 8 degrees either side of the desired ground track with certainty. On the chart, sketch a circle round each buoy corresponding to your estimate of the range of visibility. If a 15° arc plotted from the previous buoy cuts inside this circle you have a very good chance of finding it. If not, steer carefully or think again.

Radar in Fog

Radar literally sees through the fog. An operator with some experience can use radar for buoy hopping and fixing, for working up rivers and into anchorages, and numerous other inshore piloting duties. If you also have a Loran or a GPS, you can use their readings to confirm your radar fixes, which is first class navigational practice.

Fog and Electronic Fixing Aids

Just as with close quarters boathandling under sail, navigation in fog using Loran, GPS or even radar should not be undertaken without a safe way out if things begin to go wrong. If you are well offshore, you can continue with your conventional plot backed up by electronics. Inshore, however, you must be a lot more careful.

An important rule to remember is:
Never rely upon electronics to approach a tricky piece of pilotage from which you could not extricate yourself if the systems were to go down.

Using Loran C in Low Visibility

In this example, both *Hapless* and *Corker* are equipped with Loran C. When the fog comes down, *Corker* treats the situation as though its Loran was about to break down. It steers well to the safe side of the entrance, checking its position against the readout until it found the "entrance contour" with its fathometer. *Corker* then turns onto this, and punches in the position for the bell buoys as a waypoint. *Corker* finds the entrance before it was expected, but in complete safety.

1

2 When the fog comes down, *Hapless* steers straight for what she believes will be the entrance to Nantucket Harbor. Unknown to both skippers, there is a fixed Loran error causing the position to appear 150 meters to the west. The result is that *Hapless* ends up lost and confused to the west of the entrance. Had visibility been 200 meters, *Hapless* might have spotted the bell buoys, but it wasn't, so she ran aground.

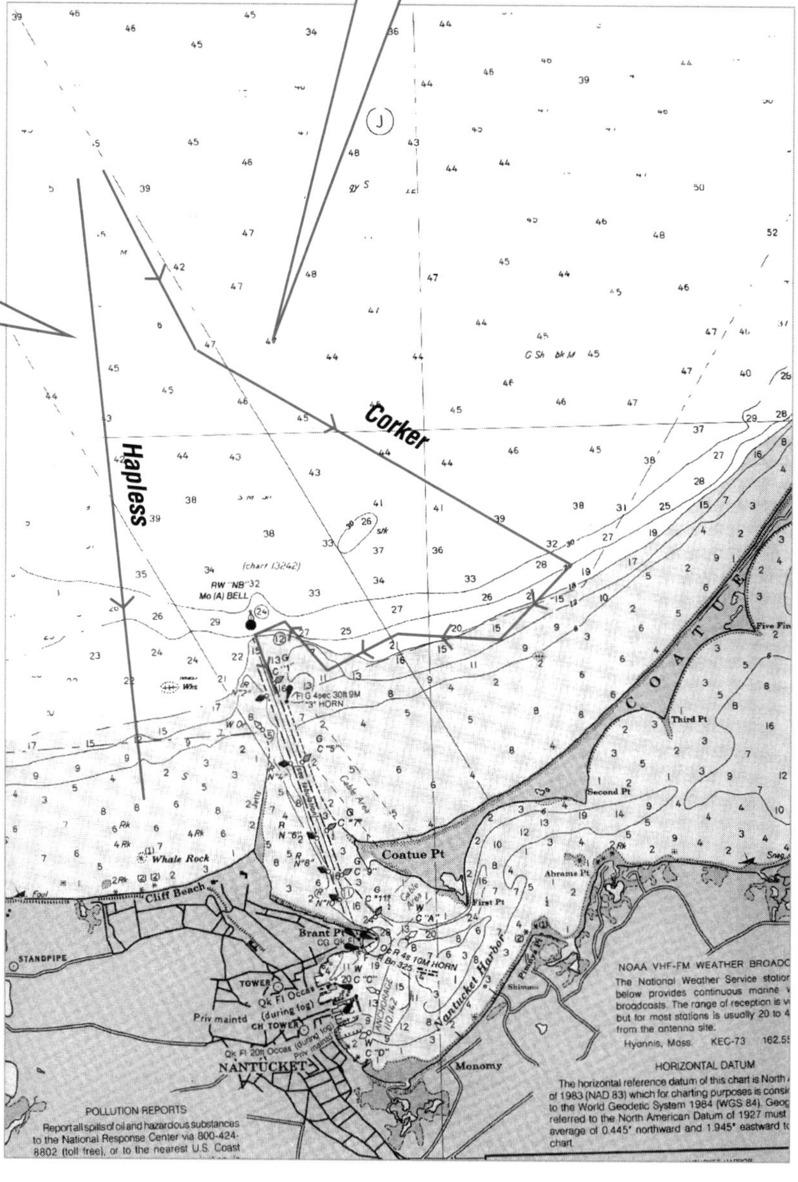

In this situation, if both boats had been fitted with GPS, both would have been OK, but had the receivers gone down, *Corker's* fail-safe system was already in place. *Hapless* would have needed to do some very quick thinking indeed.

CHAPTER 15 Planning a Passage

The successful skipper/navigator does not merely decide where to sail today, and then shove off. He or she spends a bit of time planning the proposed trip. It is this planning process that often determines the success of a passage and the happiness of the crew.

Skippers who plan their passages effectively are usually popular people to sail with. Those who don't, run into all sorts of trouble through "bad luck." The weather does not deal fairly with them, they miss crucial current changes by half an hour, or they run short of fuel or beer or both. For them, nightfall seems to come sooner than expected.

This chapter reviews a checklist of items to consider before setting out on a passage. You will not need to deal with them all every time, but each will stand you in good stead for running a happy ship.

Charts

Make sure you have all the charts you need and that they are sufficiently up-to-date for your purposes. It is all too easy to set off with a passage chart, imagining that it will be adequate, only to find yourself confronted by a maze of unknown buoys and beacons as you enter harbor at journey's end. Go through the passage in advance, and look at each part of it as though you were actually navigating it. Make sure, too, that you have any charts you may need in the event of a change of plan.

Coast Pilot Books

Have you studied the passage in the relevant Coast Pilot and/or cruising guide? These should be required reading for a first time visit to unfamiliar harbors. Work through any pilotage notes in conjunction with your charts. This will show you straight away if they have sufficient detail for the job.

Distances and Time Probabilities

Note the distance to your proposed destination. How long is it likely to take at your expected speed? Does this make sense in the context of your cruise plan? Can you complete the passage in daylight? What about a dawn landfall? If it is a longer passage, how many days and nights is it likely to take?

Passage Planning Checklist

- ☐ Review relevant charts and any updates
- ☐ Review relevant updates to Coast Pilot and/or cruising guides
- ☐ Determine distances and estimated times
- ☐ Determine viable alternatives and ports of refuge
- ☐ Identify suitable waypoints and double check their latitude and longitude
- ☐ Check short term weather forecast and developing trends
- ☐ Determine tides and currents and use them to your advantage
- ☐ Identify any tidal height considerations relevant to harbor entrances and exits
- ☐ Identify dangers and obstacles
- ☐ Note approximate basic courses
- ☐ Prepare appropriate foods

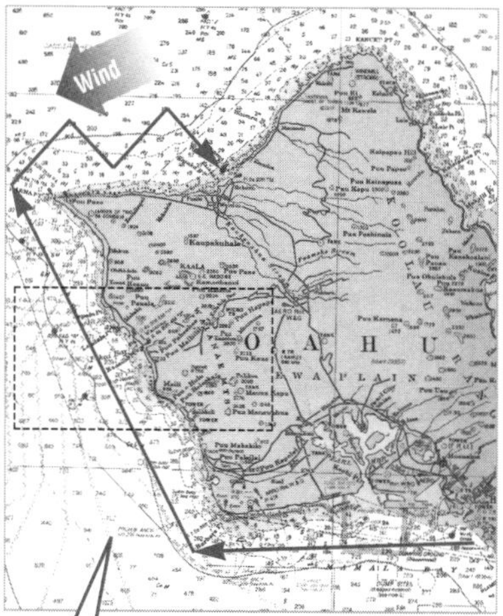

Wanderer has decided to leave Ali Wai Yacht Harbor in Honolulu (see above) and sail to Haleiwa on the north shore of Oahu. Knowing the prevailing tradewinds are easterly to northeasterly they anticipate the trip beginning with a broad reach and ending hard on the wind. In the middle of the trip there is a 20 mile section where the winds can alternate from light to heavy as they pass a number of valleys leading down out of the Waianae Mountains. It is also probable the last 10 miles of the trip will be a beat to weather in building seas. It is reassuring to know that there is a harbor of refuge at Pokai Bay, should it be necessary to lay over and take a break. *Wanderer*'s navigator has carefully checked over the chart and read the pertinent parts of the COAST PILOT and is comfortable they can safely enter Pokai Bay. If all goes well, the trip should take 18-20 hours.

Viable Alternatives/Ports of Refuge

The essence of recreational cruising is flexibility, but even for the commercial yacht skipper things do not always go according to plan. One or two viable alternatives can make a destination more attractive to the prudent planner.

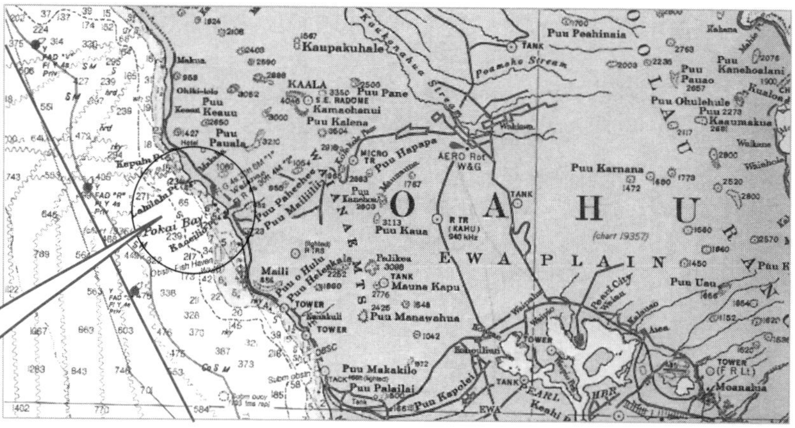

Waypoints

The best time to decide on waypoints you hope to use during a passage is before you start. Identify strategic positions on your passage chart, including your destination, and see what they look like on the detail charts. It is not uncommon to choose a waypoint that does not stand up to a closeup view because of proximity to danger. Double check the latitude and longitude of each waypoint, then punch it into your receiver's computer, looking twice at each entered waypoint before moving on to the next. Don't forget that human error, usually at the planning stage, is the most common source of poor results from electronics. It even beats power loss on the disaster scale.

Give each waypoint a name or a number in the computer and record this on the chart by the waypoint itself. If your receiver has enough memory, it may also prove useful to hold them, noting each in a waypoint log for future use.

Weather

When planning a passage of more than a day's duration, obtaining detailed medium-term weather information is a must. There is little worse than finding yourself at the wrong end of a 36-hour beat home because you failed to note that the weather front due over the weekend was going to alter the prevailing wind direction radically. Make it your business to study weather trends before you leave to be ready for any pattern which might develop.

Most importantly, pay close attention to the chapter in this book on navigational tactics and strategy. This is where it all pays off. It's also valuable to be familiar with the performance of the boat in which you will be sailing. While most modern yachts of 35 feet or over can beat to windward in all but the worst weather, some are better at it than others. If you ask an unsuitable vessel to sail its way up a hard beat, you may well end up motorsailing against the wind in a steep sea. The crew will hate it and so will you.

NAV. NOTES...

VMG (Velocity Made Good) defines a boat's speed toward an imaginary point rather than speed through the water. VMG does not take into account current. On a reach, VMG toward a destination straight ahead is often the same as the speed reading on the log. On a beat, however, VMG is usually different from speed through the water. For example, a boat making a speed of 5 knots through the water on a beat may be making three knots VMG toward a point directly upwind.

Tidal Currents

You can't change current, but you can plan around it. This usually involves taking steps to cross it at the most advantageous point, to minimize its effects when going against it by sailing in shallow water, or to use the full benefit if it is fair.

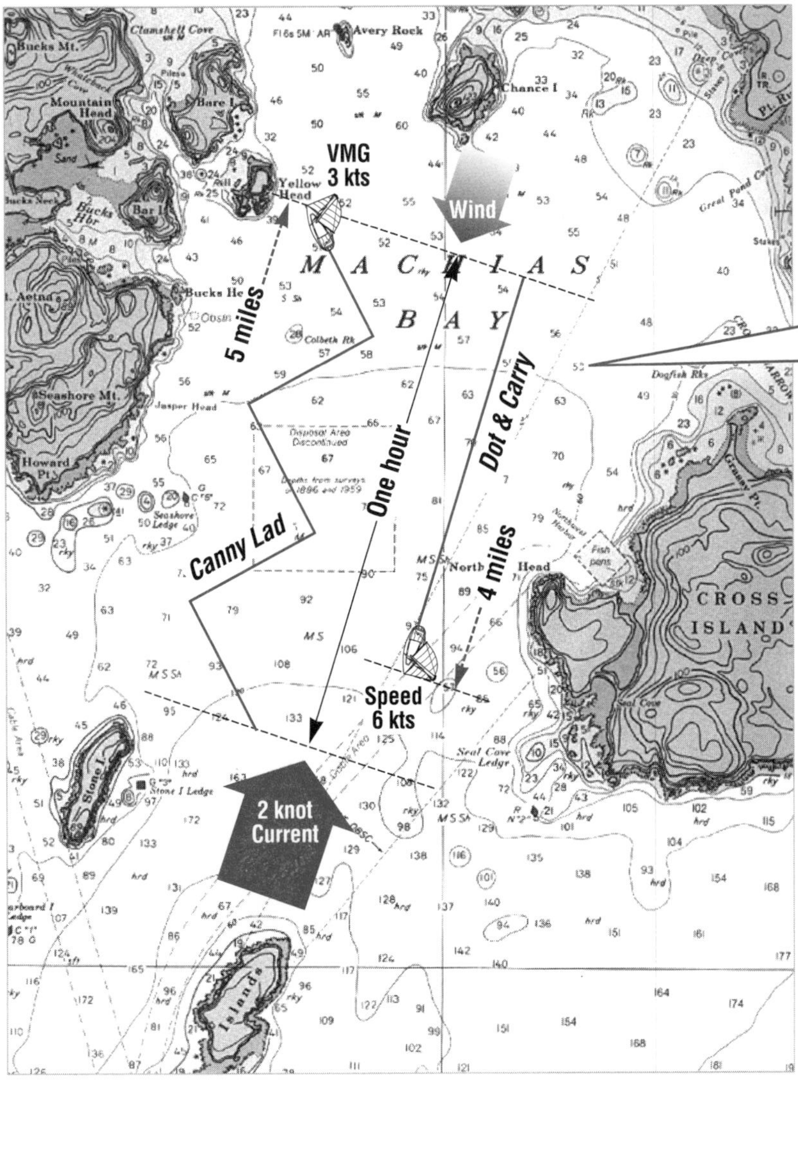

Tidal currents, if they run hard, have more impact on a passage than wind direction. *Canny Lad*, shown beating into 15 knots of wind with a 2 knot fair current, is making a good 5 knots over the ground.

3.0 knots boat VMG
+ 2.0 knots current
= 5.0 knots made good

Dot & Carry, running bravely against the current in the same wind, is managing only four.

6.0 knots boat speed
- 2.0 knots current
= 4.0 knots made good

Tidal Gates — These are places where the tide runs extremely hard. It pays handsomely to plan your passage through a tidal gate around the times of fair tide, even if it means leaving at an unsociable hour. There are numerous such places, from The Golden Gate leading into San Francisco Bay to Hell Gate on the East River leading into New York Harbor. Ignore them at your peril.

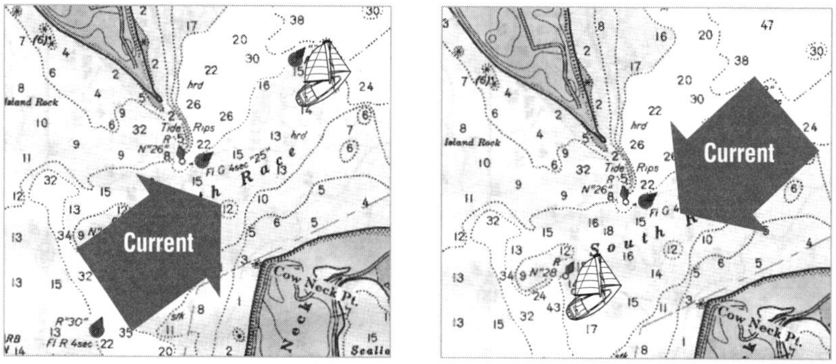

Canny Lad (left) has timed her arrival at this tidal gate with a fair tide. *Dot & Carry* (right) has missed the fair tide and is almost stationary.

When the passage is in progress, note how your speed is carrying you towards a tidal gate. If you are running late, you had better start motorsailing to make up time, or find a suitable anchorage to await the next fair tide.

Off-lying Dangers and Obstacles

Make a mental note of these before leaving. It isn't necessary to go to the extent of writing them all down, but it is well worth familiarizing yourself with them and marking clearing lines on the chart.

Traffic Separation Schemes fall into this category. Plan for them in advance. They can be particularly inconvenient when they lie across your course if you are just laying your destination.

A **Traffic Separation Scheme** is set up by the authorities to keep ships apart in areas where there is a high risk of collision.

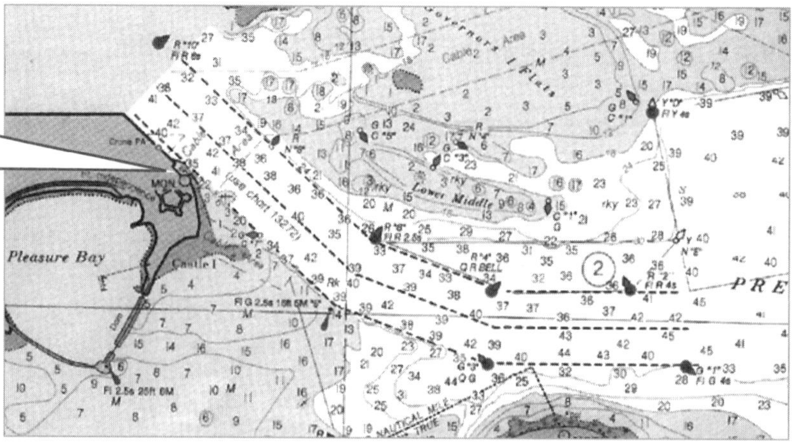

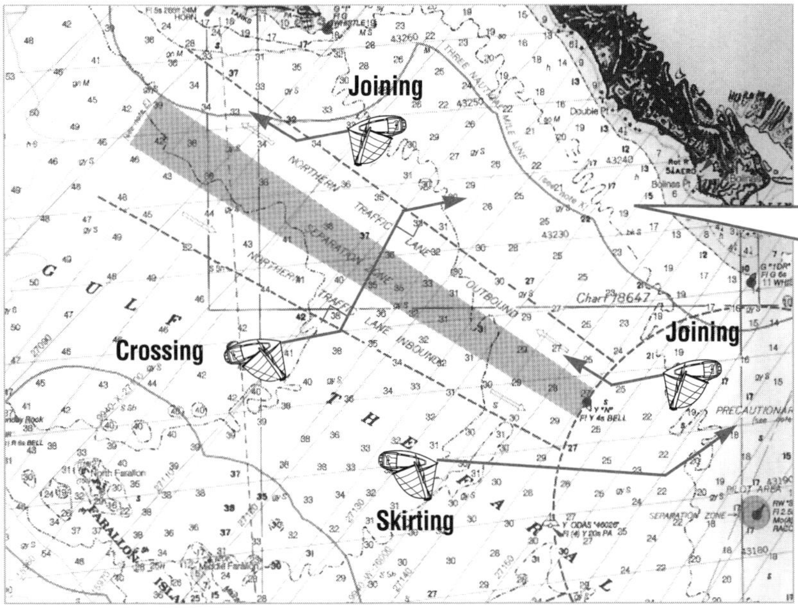

Your duties in a Traffic Separation Scheme are as follows:

▶ You should keep well clear of Traffic Separation Schemes if convenient.

▶ If you need to join the traffic in a scheme, you should do so at the ends if this is reasonable. If not, you must "go with the flow" on pain of heavy fines.

▶ When crossing a scheme, you are obliged to cross at right angles to the traffic flow at your best speed. If this means motoring, then so be it. "Right angles" means 90° relative to the traffic, not the seabed. In other words, if there is a current do not lay off for it. This is so that you will cross in the shortest possible time and will present an aspect of 90° to the shipping.

▶ The separation zone must be crossed at right angles as well. Do not resume your course or run along inside it parallel to the traffic.

▶ You can go around the edge of a Traffic Separation Scheme without being subject to the above rules.

Tidal Heights

Occasionally there will be a critical tidal "time window" for entering or leaving a harbor. If so, the hours when harbor access is available will be a critical part of your passage plan.

Basic Courses To Steer

Do not work these out in too much detail because, as we have seen, a sailing boat does not often travel straight down the rhumb line for reasons of speed, comfort, or wind angle. Nonetheless, it is worth noting an approximate heading which can be modified in the face of reality. It is optimism of the highest order to draw pre-planned lines on a chart and to announce to yourself that "this will be my course!" (See also Chapter 17, *Passage Navigation*).

Provisioning

This might sound obvious, but taking along the right food and drink can make all the difference to a trip. If rough weather is expected on an "overnighter," make a good, hearty stew before leaving and keep it in the pressure cooker, lid clamped down, waiting for its moment of glory. Don't let the bread run out, and do not be over ambitious on the gourmet front. Simple, easily prepared meals are the order of the day. Keep the fancy stuff for when the anchor is down. Do not run out of fresh water.

CHAPTER 16 Navigating in Heavy Weather

The previous chapters have all made the assumption that you are able to sit or stand at a chart table, that the chart stays on it and that you are not critically seasick. In heavy weather, none of these premises is guaranteed. The usual sources of information can also become far more difficult to use and less accurate. It is essential to realize that this will happen and to have a contingency for coping with difficult piloting weather. Here are some of the problems that gale or near gale conditions can present to the sailor in open water.

Seasickness can make every visit to the navigation station a test of character and endurance as you fight against nausea.

Motion and wetness are often extreme. Your charts can be thrown onto the cabin sole with your plotter while you are frantically hanging on to prevent yourself from following them. Even if the boat is totally tight (which cannot be assumed), the water dripping off the hood of your foul-weather gear can soak a chart in a short time if you are not careful.

Low visibility frequently accompanies gale force winds. It may not be technically foggy, but in the spume and the general dampness that pervade before a warm front arrives, you may not be able to see as much as you would like to. Even if you have clear air, spotting and identifying buoys as they dance up and down in high seas are difficult. Since your boat is also disappearing behind every other wave, counting a light sequence can be impossible until you are quite close.

Leeway increases dramatically in a big sea. Do not be surprised to experience 15° or more when reaching with a gale on the beam. A wind-driven surface current alone can account for much of this. If in doubt, always assume the worst leeway.

Distance logs become erratic as their impellers pitch in and out of the water and trailing log lines can be blown clear into the air.

Courses steered and bearings from hand bearing compasses are unlikely to be accurate. Even the best helmsman will struggle to keep a straight heading and the damping of the average compass will not cope well with extreme motion.

Personal performance suffers even if you are not seasick. The physical effort required to perform even simple duties in heavy weather is such that lethargy can begin to sap your morale. Be aware of this tendency and do what is necessary to fight it.

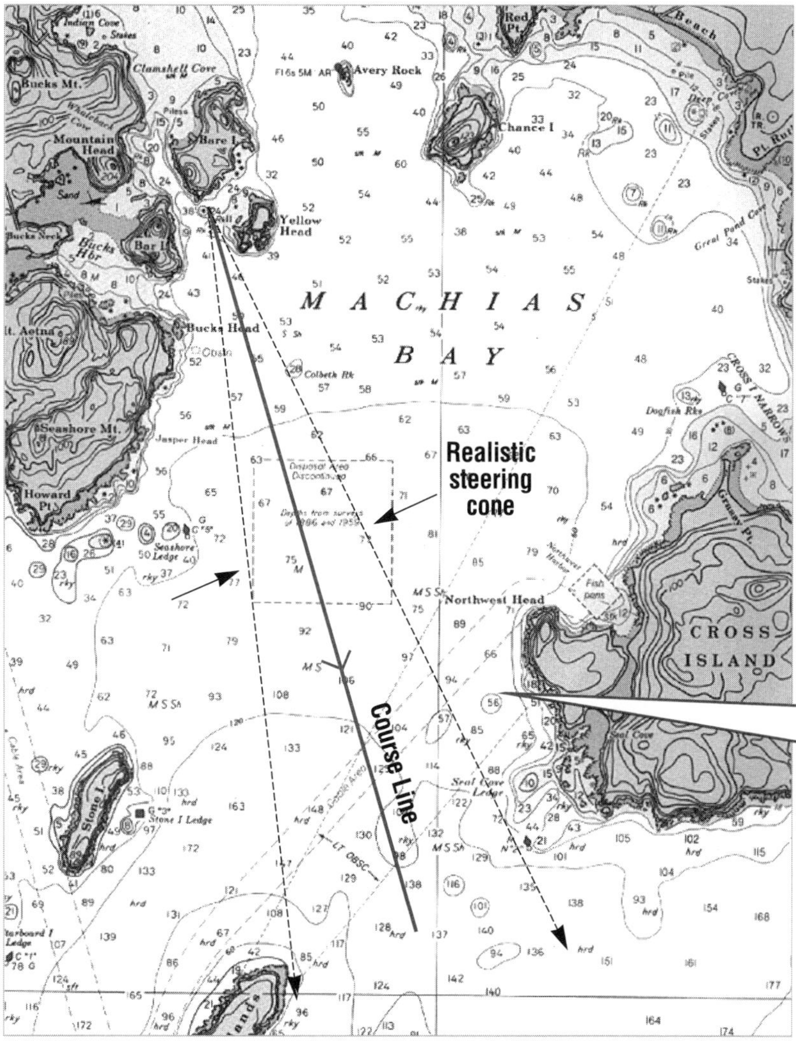

The Safety Cone

Because rough weather makes steering a straight course difficult and dramatically increases leeway, it is vital not to plot a course that will pass too close to danger. Instead of sailing to a "course line" it is wise to assume that you may end up anywhere inside a cone fanning out at 10° from either side of your departure point. If the cone passes too close to a hot spot for comfort, adjust your course and keep a good lookout, particularly if the adjustment places the outer borders of your cone near problems on the other side.

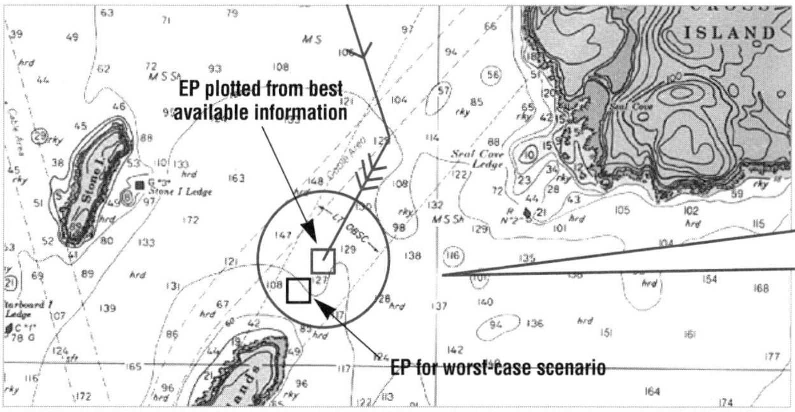

Circle of Probability

If you assume your course may be "off" by a certain number of degrees on either side, you should take this into account when plotting a DR or an EP. The result will be a *circle of probable position.* It would be unwise to hope for better. To be safe, assume you are in the part of the circle closest to danger.

Taking a Heavy Weather Fix — The difficulty of using a hand bearing compass in a big sea can make fixing your position uncertain. Use a range where you can. You can also employ the steering compass for bearings by "twitching" off course for a few seconds to aim the boat towards an object. Take full advantage of the "object abeam" source of LOP described in Chapter 10.

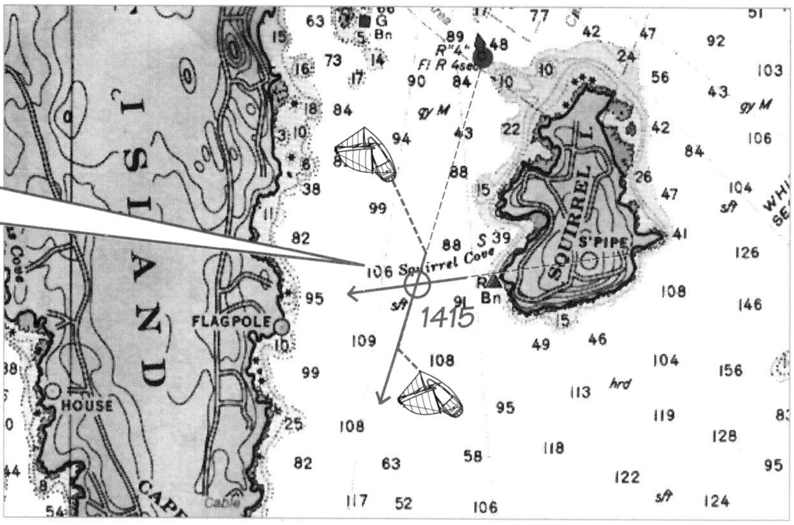

This fix has been easily achieved by steering towards a charted object to obtain a bearing at the moment the boat crosses a range that is almost abeam. It is only a two-line fix, but both LOPs are likely to be reasonably accurate.

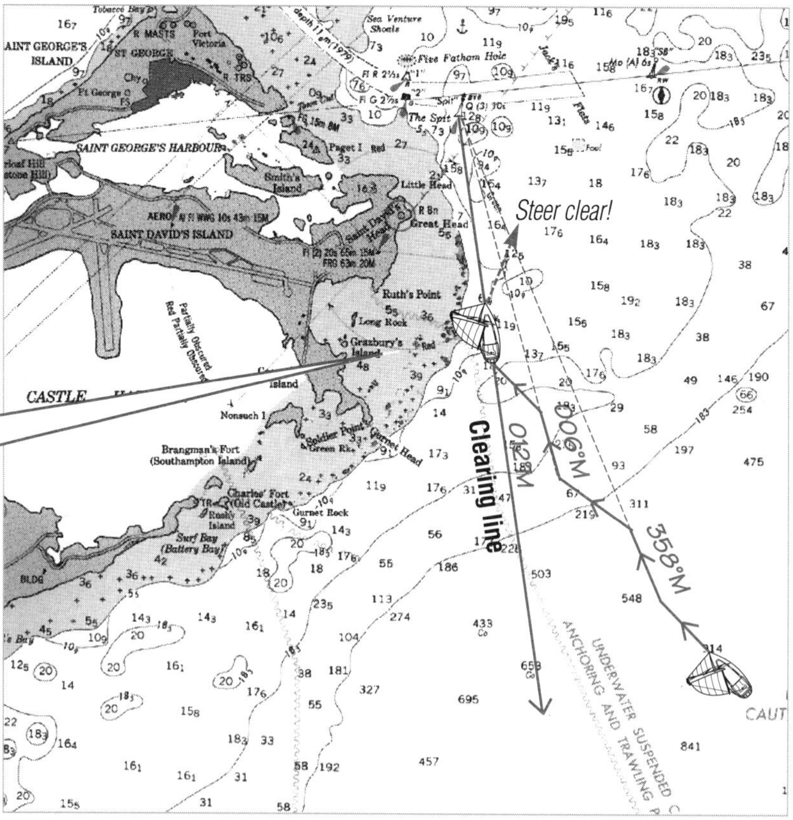

Checking a Clearing Line

If possible, use the ship's heading to check clearing lines rather than have a crew member struggle with the hand bearing compass. You know your steering compass is good (*see Chapter 5*). In this case, a series of small alterations in course to check the bearing is all that is needed to identify the time when a major alteration is required to stay clear of danger.

Use of Electronics

It is in fog and heavy weather that electronics really earn their keep. The cross-track error function of a GPS or Loran C receiver is of enormous value when a climb up the companionway stretches your inner reserves. Make full use of the machine now, but do not forget the rules. Always maintain your ship's log so that if the receiver is disabled you can immediately work up an EP. Plot your electronic positions regularly on the chart in order to make sure they form a predictable pattern. If one looks amiss, find out what is happening to you. Is it the fix that is wrong, or your estimate of where it should have been? If you have failed to keep up your log there will be no way of knowing. This principle is further expanded in the electronic navigation chapter under the heading of "Human Error."

Organization

If the weather looks like it is deteriorating, make sure your passage plan is fully up to date. You will not feel like doing more than the bare minimum when seas are running high.

Organize your chart table like a battle center. Everything must be in its place and readily at hand. All of your tools should be secure. Small items such as dividers and pencils are only too willing to disappear, and it is a shame to make oneself seasick crawling about in the bilge searching for them.

As navigator on board a yacht, you are probably also skipper. You are the most important person on board and you must keep yourself in good shape, especially if you are in danger of succumbing to seasickness. Every trip to the navigation table must therefore be for a specific purpose. Define what this is before you begin and if it is not absolutely necessary, don't do it. Now is the time to be truly realistic about your navigational needs. Make as few plots as you safely can, but *make each one count!*

CHAPTER 17 Passage Navigation

So far, we have considered the individual skills which add up to the coastal navigator's armory. Learning them is the beginning of wisdom. Nevertheless, one of the biggest problems the student skipper must face is the question of which navigational skills should be deployed in a given set of circumstances. There is a natural tendency to over-navigate, to feel that one must use every bit of knowledge one has, in order to be safe. It is critical, therefore, to work towards an ability to sum up a question quickly, decide what is required, then deal with it so that the practical business of skippering the yacht can carry on. Don't forget, as long as you are sure you are out of danger, you do not need to know the yacht's exact position all the time.

In this chapter, a real-life challenge in passage navigation is presented. It involves a decision concerning current and an arrival executed at night. Good luck and smooth sailing!

A Passage from Kaunakakai to Kaneohe Bay

Charts

Three charts are used for this passage: #19353 for the departure from Kaunakakai, #19340 for the entire passage, and #19359 for the arrival in Kaneohe Bay. Be aware that the depth scale changes from feet to fathoms and back to feet again as we change charts.

Passage Plan

The weather forecast is for relatively weak trade winds of 12 to 15 knots. We will be fairly sheltered along the south coast of Molokai, but will encounter good winds (and seas) on the beam after we pass Laau Point. We have chosen our departure time because we need to pay our bill with the harbormaster at Kaunakakai. We are comfortable arriving at Kaneohe after dark since the chart indicates plenty of well lit navigation aids. We would not want to arrive in the early evening given the large amount of coral protecting Kaneohe Bay, since the buoys could prove hard to spot against the west-setting sun.

On this trip we don't need to be concerned about reversing current — just a wind driven current that is fairly steady.

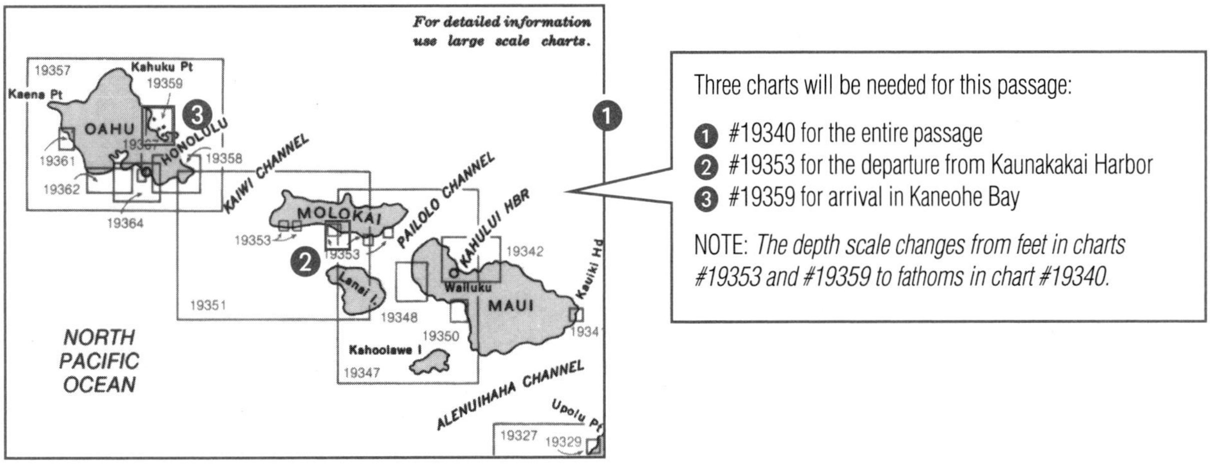
For detailed information use large scale charts.
Kahuku Pt
Kaena Pt
OAHU
HONOLULU
KAIWI CHANNEL
MOLOKAI
PAILOLO CHANNEL
KAHULUI HBR
Wailuku
MAUI
Lanai I.
Kahoolawe I
ALENUIHAHA CHANNEL
Upolu Pt
NORTH PACIFIC OCEAN
Three charts will be needed for this passage:
1 #19340 for the entire passage
2 #19353 for the departure from Kaunakakai Harbor
3 #19359 for arrival in Kaneohe Bay
NOTE: *The depth scale changes from feet in charts #19353 and #19359 to fathoms in chart #19340.*

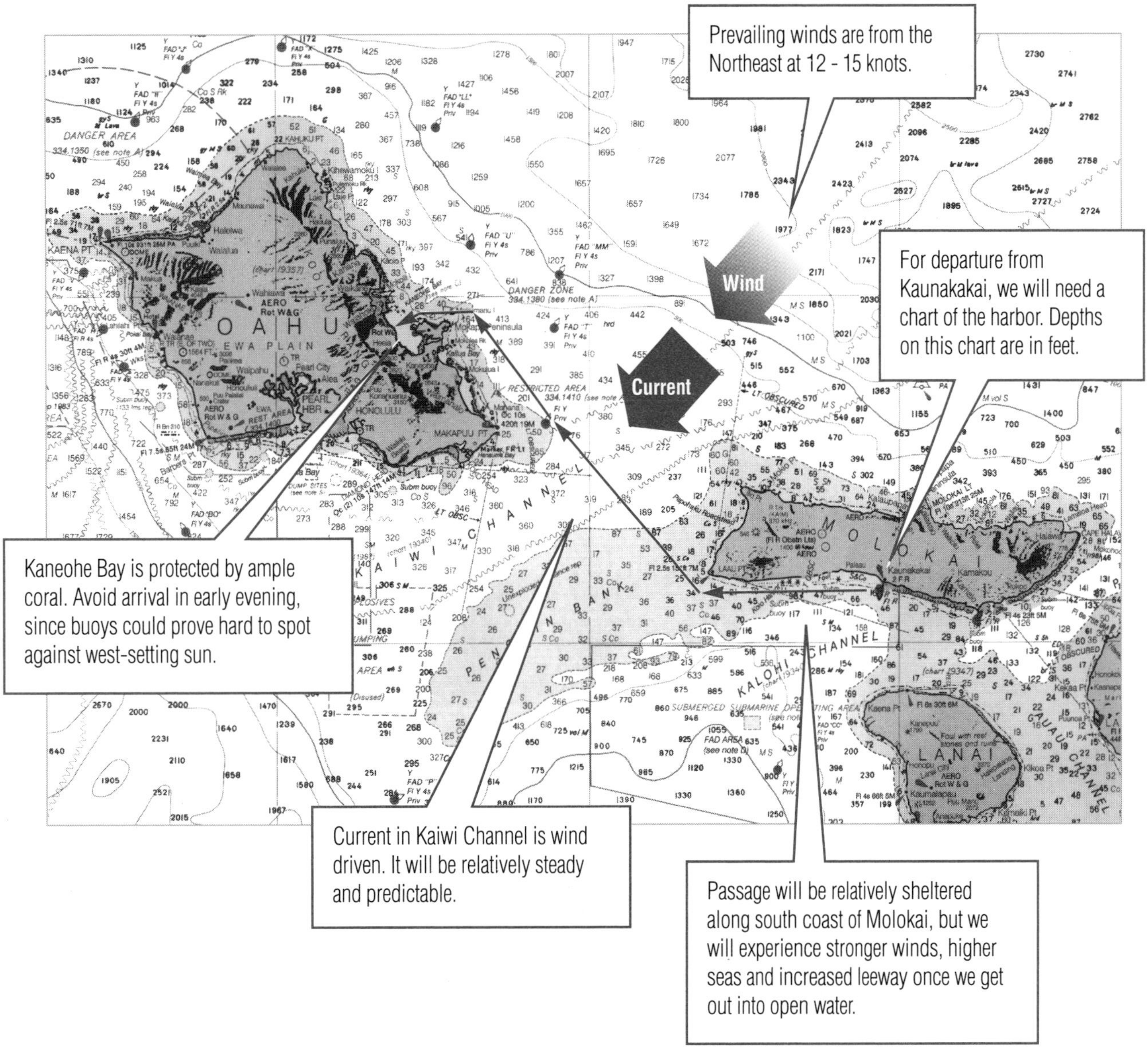
Prevailing winds are from the Northeast at 12 - 15 knots.
Wind
Current
For departure from Kaunakakai, we will need a chart of the harbor. Depths on this chart are in feet.
OAHU
MOLOKAI
LANAI
KAIWI CHANNEL
KALOHI CHANNEL
Kaneohe Bay is protected by ample coral. Avoid arrival in early evening, since buoys could prove hard to spot against west-setting sun.
Current in Kaiwi Channel is wind driven. It will be relatively steady and predictable.
Passage will be relatively sheltered along south coast of Molokai, but we will experience stronger winds, higher seas and increased leeway once we get out into open water.

Leaving Kaunakakai Harbor

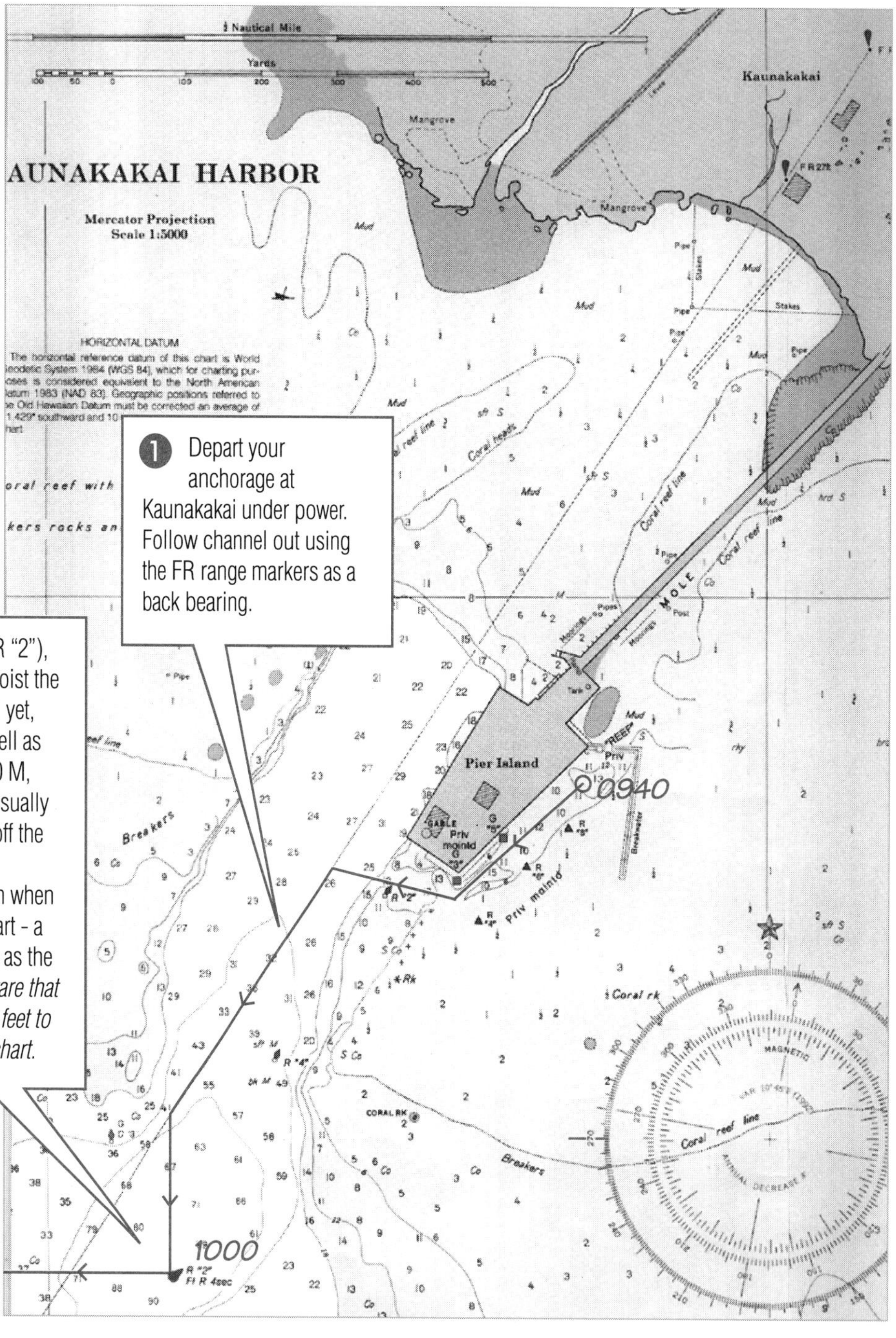

❶ Depart your anchorage at Kaunakakai under power. Follow channel out using the FR range markers as a back bearing.

❷ At the entrance buoy (R "2"), reset the trip log to 0, hoist the main (not enough wind to sail yet, but it will definitely help, as well as steady us), alter course to 260 M, and change charts, steering visually to keep Haleolono Point just off the starboard bow. We should be absolutely sure of our position when we transfer it onto the new chart - a relatively easy procedure here as the buoy is on both charts. *Be aware that the depth scale changes from feet to fathoms and feet on the new chart.*

Time	Log	Course	Weather	Remarks	Engine
0940	42.4	Various	5 NNE	Depart anchorage off Pier Island, Kaunakakai Harbor, Molokai for Heeia Kea Small Boat Harbor, Kaneohe Bay, Oahu. Following channel & range markers.	On
1000	0.0	260 M	5 NNE	Close aboard entrance buoy R "2" off Kaunakakai Harbor. Hoist mainsail. Re-zero trip log. Shift plot from Kaunakakai Harbor chart to chart #19340. N.B. Depth on this chart is in Fathoms. Set course for WP 1.	On

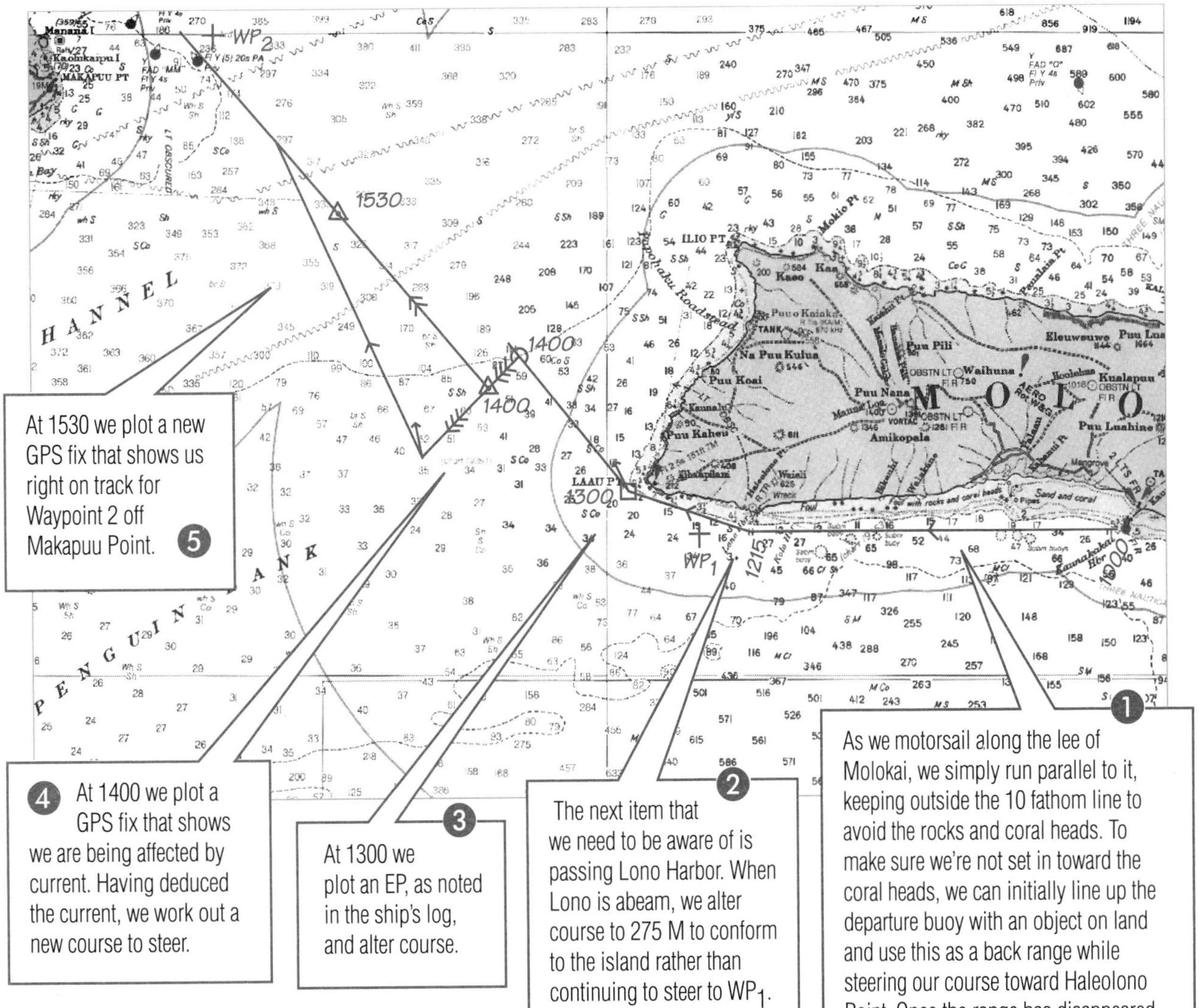
WP2
1530
1400
1400
1300
1215
1100
WP1
LAAU PT
MAKAPUU PT
At 1530 we plot a new GPS fix that shows us right on track for Waypoint 2 off Makapuu Point. 5
4 At 1400 we plot a GPS fix that shows we are being affected by current. Having deduced the current, we work out a new course to steer.
3 At 1300 we plot an EP, as noted in the ship's log, and alter course.
2 The next item that we need to be aware of is passing Lono Harbor. When Lono is abeam, we alter course to 275 M to conform to the island rather than continuing to steer to WP1. At 1250 the wind increases, allowing us to start sailing.
1 As we motorsail along the lee of Molokai, we simply run parallel to it, keeping outside the 10 fathom line to avoid the rocks and coral heads. To make sure we're not set in toward the coral heads, we can initially line up the departure buoy with an object on land and use this as a back range while steering our course toward Haleolono Point. Once the range has disappeared, we can use the fathometer or visually check the clear tropical waters.

Along the Molokai Coast and Into Kaiwi Channel

Time	Log	Course	Weather	Remarks	Engine
1215	12.4	275 M	5 NNE	Lono Harbor entrance abeam to starboard.	On
1250	15.6	275 M	15 NNE	Wind has increased in speed. Set the working jib and shut off the engine. Sailing on starboard tack with wind abaft the beam. Seas a bit lumpy, estimate we are making 5 degrees of leeway.	Off
1300	16.5	315 M	15 ENE	Abeam of Laau Pt., distance approx. 1 mile. EP on chart.	Off
1400	22.0	331 M	15 ENE	GPS Fix. Lat 21°08'.7 N by Lo 157°24'.1 W. Current has begun to affect us. Set is 223 T; Drift is 1.5 kts. Seas are getting bigger, now estimate 7 degrees leeway. Lay new course for WP 2 (Fl Y (5) 20sec) off Makapuu Pt.	Off
1530	29.3	331 M	15 ENE	GPS Fix. Lat 21°14'.5 N by Lo 157°29'.3 W	Off

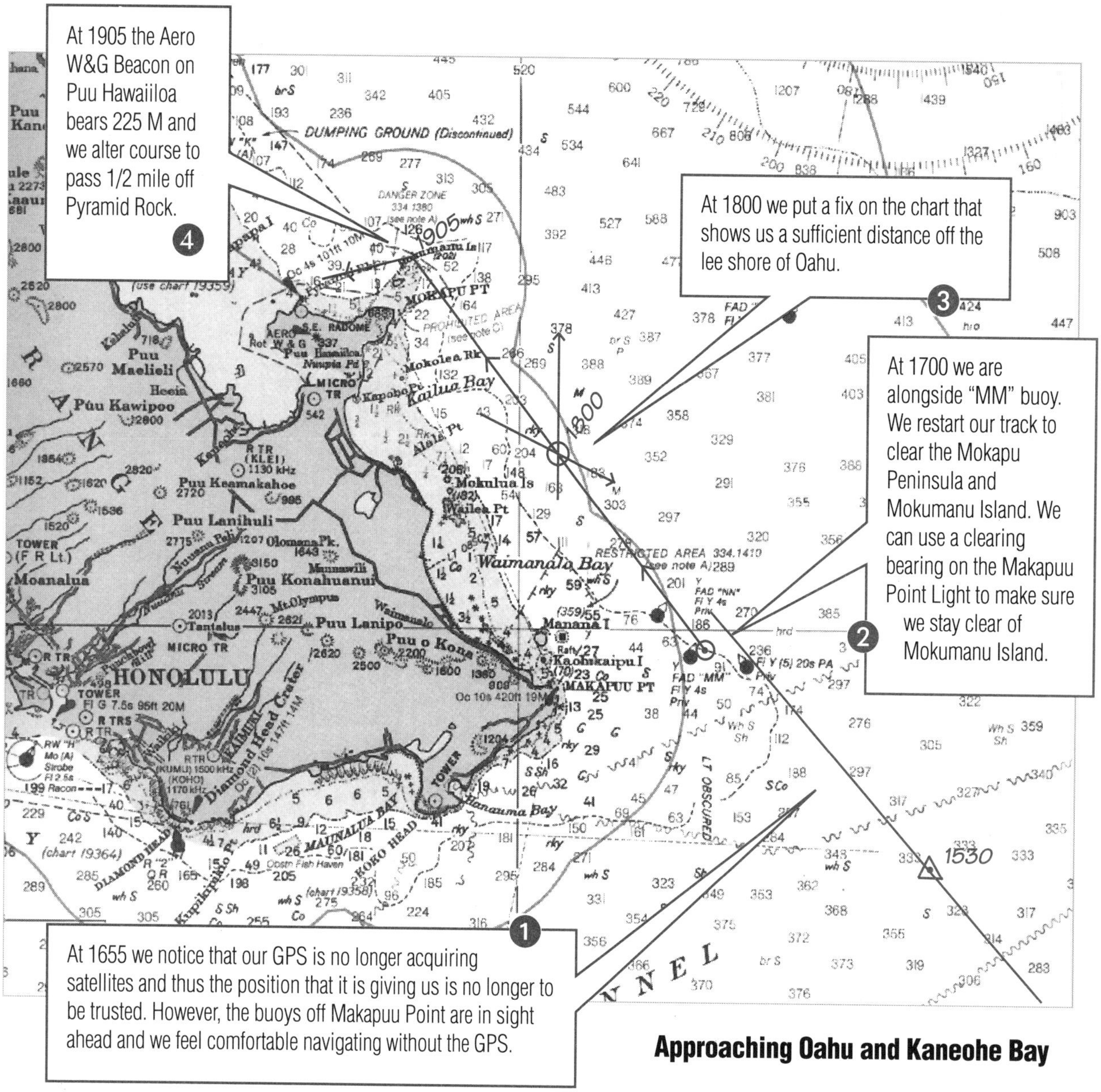
At 1905 the Aero W&G Beacon on Puu Hawaiiloa bears 225 M and we alter course to pass 1/2 mile off Pyramid Rock.
4
At 1800 we put a fix on the chart that shows us a sufficient distance off the lee shore of Oahu.
3
At 1700 we are alongside "MM" buoy. We restart our track to clear the Mokapu Peninsula and Mokumanu Island. We can use a clearing bearing on the Makapuu Point Light to make sure we stay clear of Mokumanu Island.
2
At 1655 we notice that our GPS is no longer acquiring satellites and thus the position that it is giving us is no longer to be trusted. However, the buoys off Makapuu Point are in sight ahead and we feel comfortable navigating without the GPS.
1
1905
1800
1530
DUMPING GROUND (Discontinued)
MOKAPU PT
Kailua Bay
Waimanalo Bay
HONOLULU
Diamond Head Crater
MAKAPUU PT
Hanauma Bay
KOKO HEAD
LT OBSCURED

Approaching Oahu and Kaneohe Bay

Time	Log	Course	Weather	Remarks	Engine
1655	38.0	331 M	15 ENE	GPS Just Went Down. However, the Makapuu Pt. buoys are in sight ahead.	Off
1700	38.5	317 M	15 ENE	"MM" buoy off Makapuu Pt. is close aboard, 100 yds. to port. The cruising guide for this area indicates that there is no appreciable current from here to Mokapu Pt.	Off
1731				Sunset	
1800	44.0	317 M	15 ENE	Bearing to Makapuu Pt. Light 170 M. Bearing to Aero Beacon on Mokapuu Pt. 285 M. Fix. Clearing bearing for Mokumanu Island is 153 M on Makapuu Pt. Light.	Off
1905	50.0	253 M	5 NE	Wind has died. Doused sails and we are now motoring. Aero W&G Beacon on Puu Hawaiiloa bears 225 M.	On

At 1945 we have the Aero W & G Beacon on Puu Hawaiiloa and the Occulting 4 second light on Pyramid Rock lined up as a range (in transit). R "2" and the Sampan Channel are approximately 1/2 mile ahead. We also change charts noting that on the new chart (#19359) the soundings are in feet. For this pilotage exercise entering Kaneohe Bay, we now have all hands on deck with a good "beam gun" type flashlight to pick up the reflective tape on unlighted buoys. The trick to this next piece of navigation is to have fully planned in advance all our bearings. We also want to ensure that we identify the range lights before we enter the channel. Once we have done this we must take care to identify each marker that we pass so that there is absolutely no doubt whatsoever as to our position. **1**

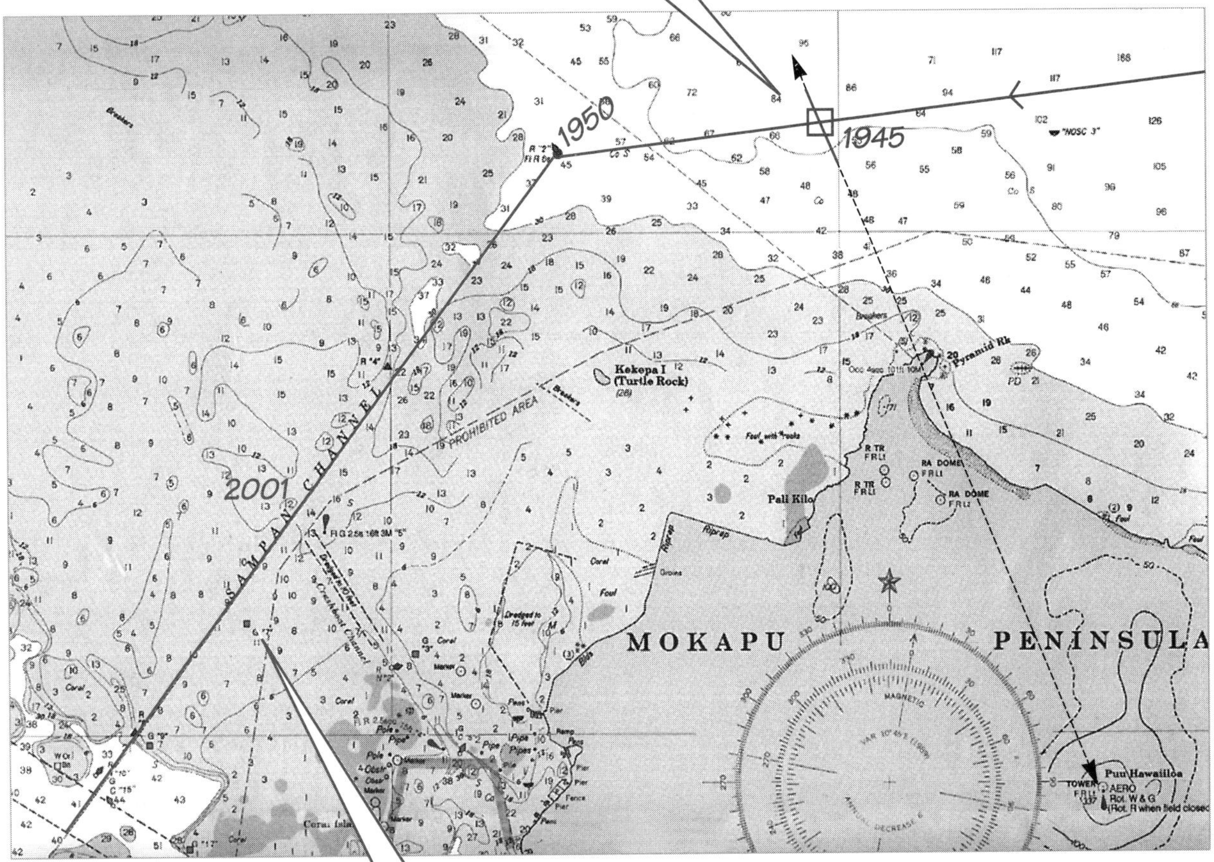

Entering Kaneohe Bay

2 At 1950, we pass close aboard buoy R "2" (Fl R 6s). Having picked up the Sampan Channel range (FR 80ft and Qk Fl R 38ft), we proceed slowly up the channel, noting the markers as we go by them, R "4" to starboard, FL G 2.5s 16ft 3M "5" to port, G "7" to port, and then between R "8" and G "9". We then come to two buoys, a nun to starboard (N "10") that is part of the Sampan Channel and a green can to port (C "15") that is part of the main ship channel that we are just beginning to enter.

Time	Log	Course	Weather	Remarks	Engine
1945	53.3	253 M	5 NE	Aero W&G Beacon and Pyramid Rock light in transit. Change to Chart #19359. N.B. Depth on chart is in feet.	On
1950	53.8	207 M	5 NE	Close aboard buoy R "2". Following Sampan Channel Range Lights down the channel.	On

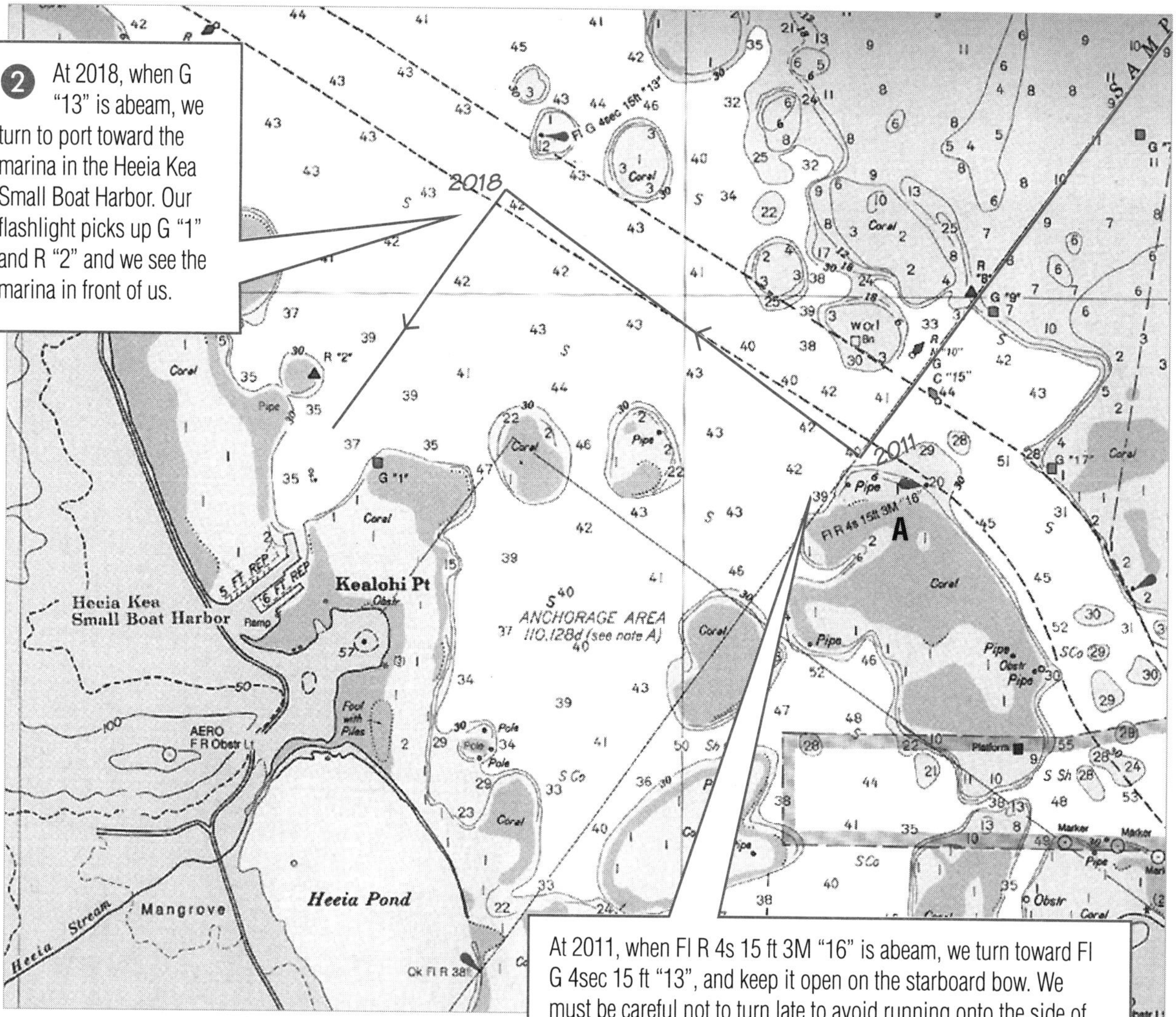

❷ At 2018, when G "13" is abeam, we turn to port toward the marina in the Heeia Kea Small Boat Harbor. Our flashlight picks up G "1" and R "2" and we see the marina in front of us.

❶ At 2011, when Fl R 4s 15 ft 3M "16" is abeam, we turn toward Fl G 4sec 15 ft "13", and keep it open on the starboard bow. We must be careful not to turn late to avoid running onto the side of the coral reef marked with a 6 foot sounding. We used the Red light abeam as our turning reference rather than a compass bearing on the more distant Green light ("13") because it is closer (*closer is better*) and we want to use the compass as little as possible during tight pilotage. To ensure we keep off the coral to starboard of our desired track after our turn, we can run a back bearing on "16."

Arriving at Heeia Kea Small Boat Harbor

Time	Log	Course	Weather	Remarks	Engine
2001	54.7	207 M	5 NE	Light G "5" is abeam.	On
2011	55.5	292 M	Zero	Light R "16" is abeam. Alter course, keeping G "13" fine on the starboard bow. Depth sounder reads 40ft.	On
2018	56.1	205 M	"	Light G "13" is abeam. Using this light as a back bearing of 025 M. Reduce speed to 3 kts.	On
2026	56.5	205 M	"	Abeam buoy G "1". We can see the dock at Heeia Kea Small Boat Harbor.	On
2035	56.8			Tied up at B pontoon.	Off

The information contained in these pages is all you need to be a successful navigator. In compiling COASTAL NAVIGATION, every effort has been made to ensure that advice given is based on sound working practice and not needlessly complicated by theory. Every aspect of the subject has been reviewed and discussed both ashore and on the water by some of the most experienced navigation instructors in the U.S. The result is a system of piloting which reflects the reality of how top cruising skippers actually operate on the water in fair weather and foul.

Those who have consulted on this work have spent years at sea, and their dedication has opened a unique door for you, the student. From day one, you will learn piloting techniques as they take place on the water, not in the classroom. You can study for hours about the characteristics of lighted buoys, but it will avail you little until you have strained your eyes into the night to identify a vital buoy in a tossing sea.

The final and most important advice of all, therefore, is to pack your seabag, step aboard and go. Piloting will burst into life with the tang of salt air, the infinite challenge of nature and the mystery of the distant horizon.

Tom Cunliffe

Chartwork & Ship's Log Entry Examples

Example #1: Logging a Course

This example shows how to log our course from Green buoy "1" off Chimney Rock to another buoy over 14 miles away in an east-southeast direction. We are estimating that our vessel is making 5 degrees of leeway while sailing close-hauled on starboard tack in a 20-knot southerly wind. We depart the buoy at 1215 with the intention of making good a course of 105 M (there is no current). At that time our distance log read 34.5.

Below is the log entry that we would have made as we passed the green buoy with the plot we would have laid out on the chart at the same time.

> Note that our log entry for 1215 records the course our helmsman is steering as 110 M. This is done so that there is no possibility of confusion at a later time over whether or not this is the course plotted in degrees True on our chart or that given to the helmsman as a magnetic course to steer. Also note that in the "Remarks" column we have recorded our fix at this time, as well as our estimate of the amount of leeway that we were allowing for determining our course to steer. In addition, we have noted that there is no known current that we must deal with.

Time	Log	Course	Weather	Remarks	Engine
1215	34.5	110 M	20 S	Close aboard buoy G "1" off Chimney Rock, Drakes Bay. Estimate 5 degrees leeway. No known current.	Off

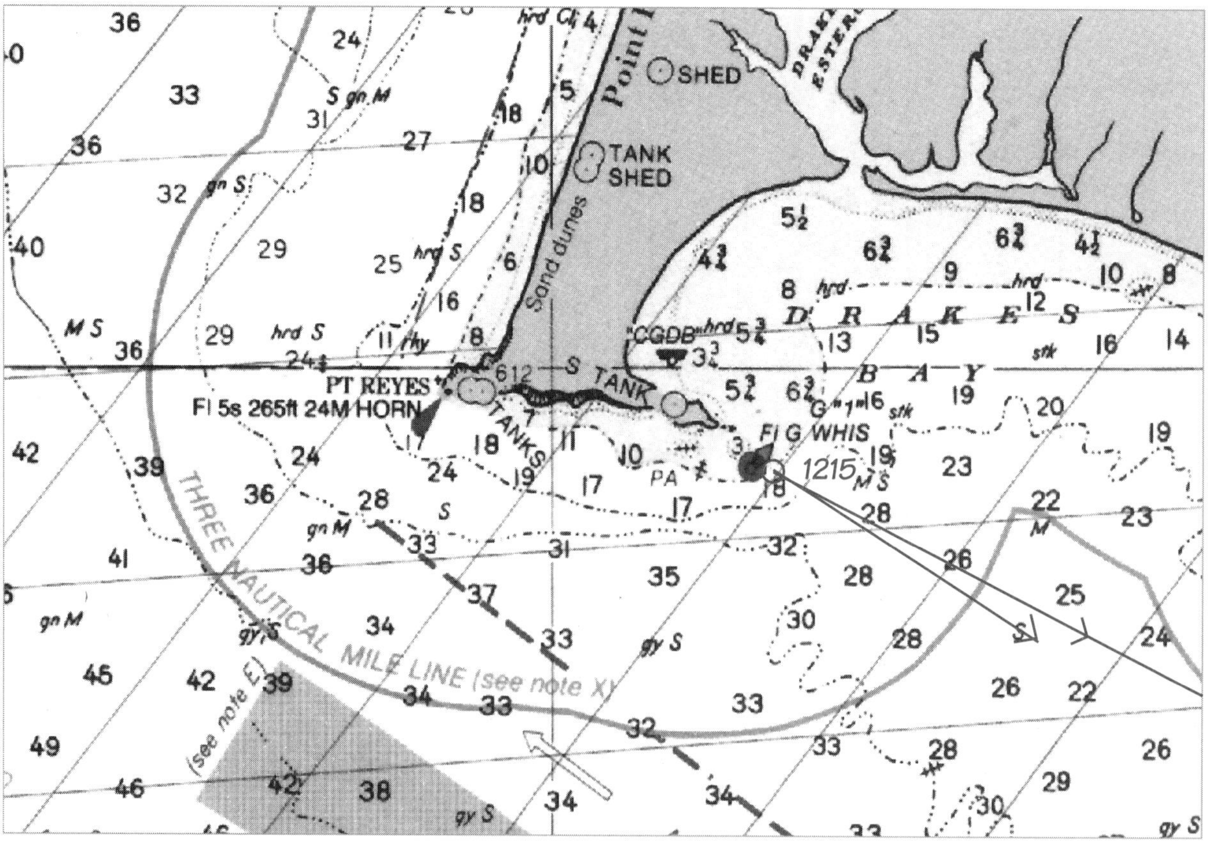

Example #2: Plotting and Logging a Running Fix

This is an example of how you should handle a running fix, showing both the various ship's log entries as well as the necessary chartwork.

Your vessel is proceeding southward on a course of 168 M. At 1450 you obtain a hand bearing compass reading on the Point Reyes Lighthouse of 105 degrees at which time your log reads 56.4 and your depth sounder indicates 28 fathoms. At 1510 you take another reading on the light, this time it is 040 degrees and the log reads 58.4 while the depth is 34 fathoms.

Below are the log entries that you should have made as you worked out your position as plotted on the chart.

Time	Log	Course	Weather	Remarks	Engine
1450	56.4	168 M	15 NW	Pt. Reyes Lt. bears 105 M. Depth 28 fathoms.	Off
1510	58.4	168 M	15 NW	Pt. Reyes Lt. bears 040 M. Depth 34 fathoms. Running fix on chart. Clear visibility.	Off

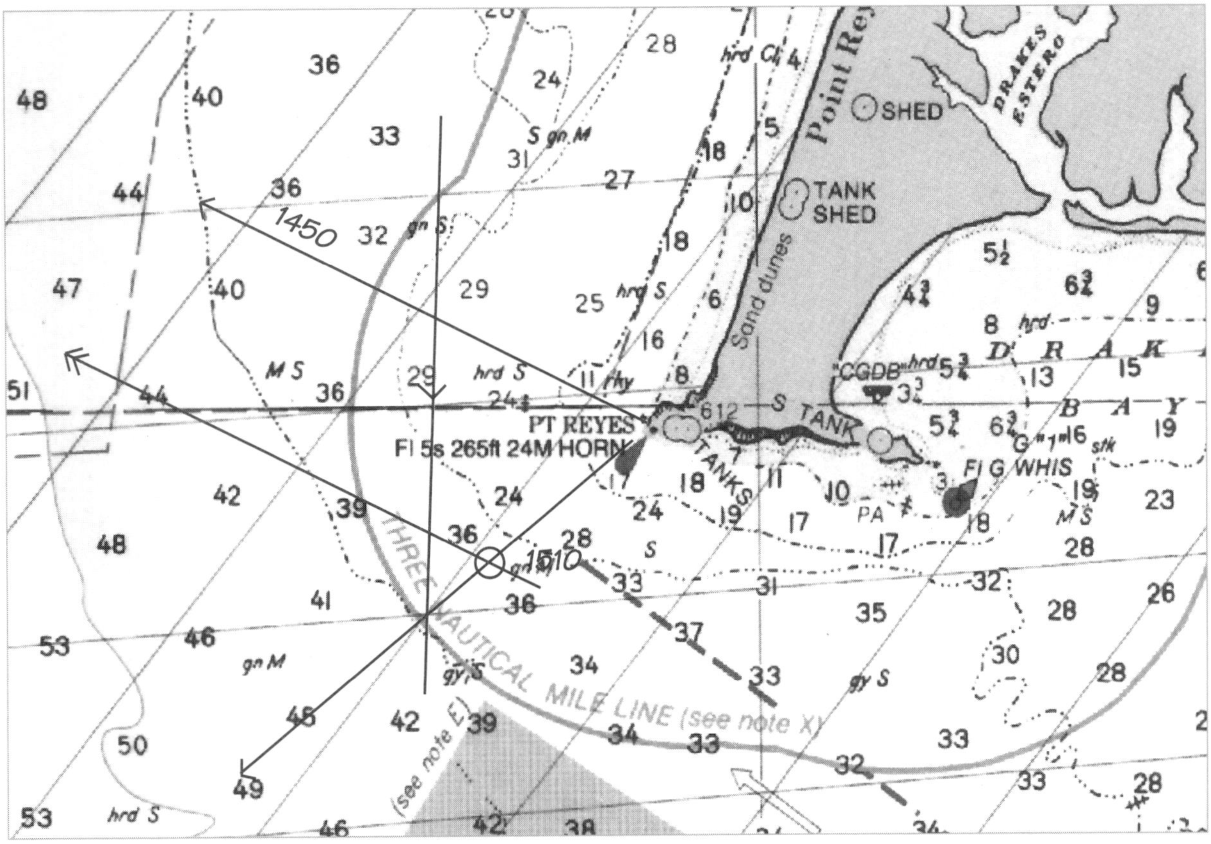

MASSACHUSETTS continued	POSITION Latitude North	Longitude West	DIFFERENCES Time High h m	Time Low h m	Height* High ft	Height* Low ft	RANGE Spring ft
Smith Point, north side	41°17'	70°14'	+0:48	–0:30	*0.16	*0.16	1.9
				ON NEWPORT			
Miacomet Rip	41°14'	70°06'	+0:15	+0:50	*0.49	*0.49	2.0
				ON NEWPORT			
Martha's Vineyard							
Wasque Point, Chappaquiddick Is.	41°22'	70°27'	+2:02	+3:20	*0.31	*0.31	1.4
Off Jobs Neck Pond	41°21'	70°35'	+0:01	+0:22	*0.77	*0.77	3.2
Off Chilmark Pond	41°20'	70°43'	–0:16	+0:04	*0.82	*0.82	3.5
Squibnocket Point	41°19'	70°46'	–0:45	–0:02	*0.82	*0.82	3.7
Nomans Land	41°16'	70°49'	–0:19	+0:18	*0.85	*0.85	3.6
Gay Head	41°21'	70°50'	–0:06	+0:45	*0.82	*0.82	3.5
Menemsha Bight	41°21'	70°46'	+0:02	+0:37	*0.77	*0.77	3.4
Cedar Tree Neck	41°26'	70°42'	+0:10	+1:32	*0.62	*0.62	2.8
Off Lake Tashmoo	41°28'	70°38'	+1:08	+2:11	*0.60	*0.60	2.5
				ON BOSTON			
West Chop	41°29'	70°36'	+0:18	–0:29	*0.15	*0.15	1.7
Vineyard Haven	41°27'	70°36'	+0:27	+0:01	*0.18	*0.18	2.0
East Chop	41°28'	70°34'	+0:29	–0:12	*0.18	*0.18	2.0
Oak Bluffs	41°27'	70°33'	+0:32	–0:12	*0.18	*0.18	2.0
Edgartown	41°23'	70°31'	+0:57	+0:18	*0.20	*0.20	2.3
Cape Poge, Chappaquiddick Island	41°25'	70°27'	+0:46	+0:04	*0.23	*0.23	2.6
Vineyard Sound							
				ON NEWPORT			
Nobska Point	41°31'	70°39'	+0:41	+2:05	*0.43	*0.43	1.9
Woods Hole							
Little Harbor..Woods Hole	41°31'	70°40'	+0:32	+2:21	*0.40	*0.40	1.8
Oceanographic Institution	41°32'	70°40'	+0:22	+1:59	*0.52	*0.50	2.3
Uncatena Island (south side)	41°31'	70°42'	+0:12	+0:22	*1.02	*1.02	4.5
Tarpaulin Cove	41°28'	70°46'	+0:11	+1:23	*0.54	*0.54	2.4
Quicks Hole							
South side	41°26'	70°51'	–0:10	+0:09	*0.71	*0.71	3.1
Middle	41°27'	70°51'	0:00	+0:10	*0.85	*0.85	3.7
North side	41°27'	70°51'	–0:08	–0:08	*0.99	*0.99	4.4
Buzzards Bay							
Cuttyhunk Pond entrance	41°25'	70°55'	+0:01	+0:01	*0.97	*0.97	4.2
Penikese Island	41°27'	70°55'	–0:17	–0:16	*0.97	*0.97	4.2
Kettle Cove	41°29'	70°47'	+0:09	+0:02	*1.08	*1.08	4.7
Chappaquoit Point, West Falmouth Hbr	41°36'	70°39'	+0:10	+0:20	*1.10	*1.07	4.9
West Falmouth Harbor	41°36'	70°39'	+0:21	+0:18	*1.14	*1.14	5.0
Barlows Landing, Pocasset Harbor	41°41'	70°38'	+0:24	+0:18	*1.14	*1.14	5.0
Abiels Ledge	41°42'	70°40'	+0:11	+0:16	*1.11	*1.11	4.9
Monument Beach	41°43'	70°37'	+0:23	+0:18	*1.14	*1.14	5.0
Cape Cod Canal, RR. bridge	41°44'	70°37'	+1:15	+2:47	*0.99	*0.99	4.1
Great Hill	41°43'	70°43'	+0:12	+0:11	*1.15	*1.21	5.0
Wareham, Wareham River	41°45'	70°43'	+0:22	+0:16	*1.16	*1.16	5.1
Bird Island	41°40'	70°43'	+0:05	–0:02	*1.19	*1.19	5.2
Marion, Sippican Harbor	41°42'	70°46'	+0:10	+0:12	*1.13	*1.29	4.9
Mattapoisett, Mattapoisett Harbor	41°39'	70°49'	+0:11	+0:20	*1.09	*1.00	4.8
West Island (west side)	41°36'	70°50'	+0:09	+0:08	*1.05	*1.05	4.6
Clarks Point	41°36'	70°54'	+0:14	+0:24	*1.06	*1.00	4.5

* Ratio

TIME OF HIGH WATER

Time figures shown are the *average* differences throughout the year. Rise in feet is mean range. (Low Water times are given *only* when they vary more than 20 min. from High Water times.)

	High	Low	H.M.			Rise in feet
Martha's Vineyard						
Cape Pogue	**high 0 45** *after*,	**low**	same as		BOSTON	2.2
Edgartown	**high 1 00** *after*,	**low**	0 15	*after*	"	2.0
Oak Bluffs	**high 0 30** *after*,	**low**	0 15	*before*	"	1.7
East Chop	**high 0 25** *after*,	**low**	0 15	"	"	1.7
Vineyard Haven	**high 0 25** *after*,	**low**	0 05	"	"	1.7
West Chop	**high 0 15** *after*,	**low**	0 30	"	"	1.4
Lake Tashmoo (inside)			2 30	*before*	"	2.0
Cedar Tree Neck	**high 0 15** *after*,	**low**	1 35	*after*	NEWPORT	2.3
Menemsha	**high 0 05** *after*,	**low**	0 40	"	"	2.7
Gay Head	**high** same as,	**low**	0 50	"	"	3.0
Squibnocket Point	**high 0 40** *before*,	**low**	0 05	"	"	2.9
Wasque Point	**high 2 05** *after*,	**low**	3 25	"	"	1.1
Nomans Island	**high 0 15** *before*,	**low**	0 25	"	"	3.0
Vineyard Sound North Side						
Little Hbr., Wds.Hole	**high 0 35** *after*,	**low**	2 25	*after*	NEWPORT	1.3
Oceanographic Inst.	**high 0 27** *after*,	**low**	2 00	"	"	2.0
Tarpaulin Cove	**high 0 15** *after*,	**low**	1 30	"	"	1.9
Quicks's Hole, S.side	**high 0 10** *before*,	**low**	0 15	"	"	2.5
Buzzards Bay						
Cuttyhunk, Pond Entrance			same as		NEWPORT	3.4
Penikese Island			0 15	*before*	"	3.4
W. Falmouth Harbor			0 25	*after*	"	4.0
Monument Beach			0 25	"	"	4.0
Pocasset Harbor			0 25	"	"	4.0
Wareham			0 25	"	"	4.1
Bird Island			0 10	"	"	4.2
Marion			0 10	"	"	4.0
Mattapoisett			0 15	"	"	3.9
New Bedford			0 10	"	"	3.7
South Dartmouth			0 30	"	"	3.7
Dumpling Rocks			same as		"	3.8
Westport Harbor	**high 0 10** *after*,	**low**	0 40	"	"	3.0
RHODE ISLAND & MASS. Narragansett Bay						
Sakonnet			0 10	*before*	NEWPORT	3.1
Tiverton			0 20	*after*	"	3.8
Beavertail			same as		"	3.5
Prudence Island			0 10	*after*	"	3.9
Bristol Point			0 20	"	"	4.1
Fall River			0 30	"	"	4.4
Taunton	**high 1 10** *after*,	**low**	2 25	*after*	"	2.8
Warren			0 15	"	"	4.6
Providence			0 10	"	"	4.6
Pawtucket			0 20	"	"	4.5
East Greenwich			0 15	"	"	4.0
Wickford			0 10	"	"	3.8
Narragansett Pier			0 10	*before*	"	3.2
RHODE ISLAND, Outer Coast						
Pt. Judith Harbor	**high 0 05** *before*,	**low**	0 20	*after*	NEWPORT	3.1
Great Salt Pond, Block Is.			same as		"	2.6
Watch Hill	**high 0 45** *after*,	**low**	1 20	"	"	2.5

BOSTON Tables, p. 12-17
NEWPORT Tables, p. 64-69

When a high tide exceeds av. ht., the *following* low tide will be lower than av.

TABLE 3.—SPEED OF CURRENT AT ANY TIME

EXPLANATION

Though the predictions in this publication give only the slacks and maximum currents, the speed of the current at any intermediate time can be obtained approximately by the use of this table. Directions for its use are given below the table.

Before using the table for a place listed in Table 2, the predictions for the day in question should be first obtained by means of the differences and ratios given in Table 2.

The examples below follow the numbered steps in the directions.

Example 1.—Find the speed of the current in The Race at 6:00 on a day when the predictions which immediately precede and follow 6:00 are as follows:

(1)	Slack Water	Maximum (Flood)	
	Time	Time	Speed
	4:18	7:36	3.2 knots

Directions under the table indicate Table A is to be used for this station.

(2) Interval between slack and maximum flood is 7:36 – 4:18 = 3^h18^m. Column heading nearest to 3^h18^m is 3^h20^m.

(3) Interval between slack and time desired is 6:00 – 4:18 = 1^h42^m. Line labeled 1^h40^m is nearest to 1^h42^m.

(4) Factor in column 3^h20^m and on line 1^h40^m is 0.7. The above flood speed of 3.2 knots multiplied by 0.7 gives a flood speed of 2.24 knots (or 2.2 knots, since one decimal is sufficient) for the time desired.

Example 2.—Find the speed of the current in the Harlem River at Broadway Bridge at 16:30 on a day when the predictions (obtained using the difference and ratio in table 2) which immediately precede and follow 16:30 are as follows:

(1)	Maximum (Ebb)		Slack Water
	Time	Speed	Time
	13:49	2.5 knots	17:25

Directions under the table indicate Table B is to be used, since this station in Table 2 is referred to Hell Gate.

(2) Interval between slack and maximum ebb is 17:25 – 13:49 = 3^h36^m. Hence, use column headed 3^h40^m.

(3) Interval between slack and time desired is 17:25 – 16:30 = 0^h55^m. Hence, use line labeled 1^h00^m.

(4) Factor in column 3^h40^m and on line 1^h00^m is 0.5. The above ebb speed of 2.5 knots multiplied by 0.5 gives an ebb speed of 1.2 knots for the desired time.

When the interval between slack and maximum current is greater than 5^h40^m, enter the table with one-half the interval between slack and maximum current and one-half the interval between slack and the desired time and use the factor thus found.

TABLE 3.—SPEED OF CURRENT AT ANY TIME

TABLE A

Interval between slack and desired time	Interval between slack and maximum current													
	h. m. 1 20	h. m. 1 40	h. m. 2 00	h. m. 2 20	h. m. 2 40	h. m. 3 00	h.m. 3 20	h.m. 3 40	h.m. 4 00	h.m. 4 20	h.m. 4 40	h.m. 5 00	h.m. 5 20	h.m. 5 40
h. m.	ft.	ft.	ft.	ft.	ft.	ft.	ft.	ft.	ft.	ft.	ft.	ft.	ft.	ft.
0 20	0.4	0.3	0.3	0.2	0.2	0.2	0.2	0.1	0.1	0.1	0.1	0.1	0.1	0.1
0 40	0.7	0.6	0.5	0.4	0.4	0.3	0.3	0.3	0.3	0.2	0.2	0.2	0.2	0.2
1 00	0.9	0.8	0.7	0.6	0.6	0.5	0.5	0.4	0.4	0.4	0.3	0.3	0.3	0.3
1 20	1.0	1.0	0.9	0.8	0.7	0.6	0.6	0.5	0.5	0.5	0.4	0.4	0.4	0.4
1 40	----	1.0	1.0	0.9	0.8	0.8	0.7	0.7	0.6	0.6	0.5	0.5	0.5	0.4
2 00	----	----	1.0	1.0	0.9	0.9	0.8	0.8	0.7	0.7	0.6	0.6	0.6	0.5
2 20	----	----	----	1.0	1.0	0.9	0.9	0.8	0.8	0.7	0.7	0.7	0.6	0.6
2 40	----	----	----	----	1.0	1.0	1.0	0.9	0.9	0.8	0.8	0.7	0.7	0.7
3 00	----	----	----	----	----	1.0	1.0	1.0	0.9	0.9	0.8	0.8	0.8	0.7
3 20	----	----	----	----	----	----	1.0	1.0	1.0	0.9	0.9	0.9	0.8	0.8
3 40	----	----	----	----	----	----	----	1.0	1.0	1.0	0.9	0.9	0.9	0.9
4 00	----	----	----	----	----	----	----	----	1.0	1.0	1.0	1.0	0.9	0.9
4 20	----	----	----	----	----	----	----	----	----	1.0	1.0	1.0	1.0	0.9
4 40	----	----	----	----	----	----	----	----	----	----	1.0	1.0	1.0	1.0
5 00	----	----	----	----	----	----	----	----	----	----	----	1.0	1.0	1.0
5 20	----	----	----	----	----	----	----	----	----	----	----	----	1.0	1.0
5 40	----	----	----	----	----	----	----	----	----	----	----	----	----	1.0.

TABLE B

Interval between slack and desired time	Interval between slack and maximum current													
	h. m. 1 20	h. m. 1 40	h. m. 2 00	h. m. 2 20	h. m. 2 40	h. m. 3 00	h. m. 3 20	h. m. 3 40	h. m. 4 00	h. m. 4 20	h. m. 4 40	h. m. 5 00	h. m. 5 20	h. m. 5 40
h. m.	ft.	ft.	ft.	ft.	ft.	ft.	ft.	ft.	ft.	ft.	ft.	ft.	ft.	ft.
0 20	0.5	0.4	0.4	0.5	0.3	0.3	0.3	0.3	0.2	0.2	0.2	0.2	0.2	0.2
0 40	0.8	0.7	0.6	0.5	0.5	0.5	0.4	0.4	0.4	0.4	0.3	0.3	0.3	0.3
1 00	0.9	0.8	0.8	0.7	0.7	0.6	0.6	0.5	0.5	0.5	0.4	0.4	0.4	0.4
1 20	1.0	1.0	0.9	0.8	0.8	0.7	0.7	0.6	0.6	0.6	0.5	0.5	0.5	0.5
1 40	----	1.0	1.0	0.9	0.9	0.8	0.8	0.7	0.7	0.7	0.6	0.6	0.6	0.6
2 00	----	----	1.0	1.0	0.9	0.9	0.9	0.8	0.8	0.7	0.7	0.7	0.7	0.6
2 20	----	----	----	1.0	1.0	1.0	0.9	0.9	0.8	0.8	0.8	0.7	0.7	0.7
2 40	----	----	----	----	1.0	1.0	1.0	0.9	0.9	0.9	0.8	0.8	0.8	0.7
3 00	----	----	----	----	----	1.0	1.0	1.0	0.9	0.9	0.9	0.9	0.8	0.8
3 20	----	----	----	----	----	----	1.0	1.0	1.0	1.0	0.9	0.9	0.9	0.9
3 40	----	----	----	----	----	----	----	1.0	1.0	1.0	1.0	0.9	0.9	0.9
4 00	----	----	----	----	----	----	----	----	1.0	1.0	1.0	1.0	0.9	0.9
4 20	----	----	----	----	----	----	----	----	----	1.0	1.0	1.0	1.0	0.9
4 40	----	----	----	----	----	----	----	----	----	----	1.0	1.0	1.0	1.0
5 00	----	----	----	----	----	----	----	----	----	----	----	1.0	1.0	1.0
5 20	----	----	----	----	----	----	----	----	----	----	----	----	1.0	1.0
5 40	----	----	----	----	----	----	----	----	----	----	----	----	----	1.0

Use table A for all places except those listed below for table B.
Use table B for Deception Pass, Seymour Narrows, Sergius Narrows, Isanotski Strait. and all stations in table 2 which are referred to these points.

1. From predictions find the time of slack water and the time and velocity of maximum current (flood or ebb), one of which is immediately before and the other after the time for which the velocity is desired.
2. Find the interval of time between the above slack and maximum current, and enter the top of table A or B with the interval which most nearly agrees with this value.
3. Find the interval of time between the above slack and the time desired, and enter the side of table A or B with the interval which most nearly agrees with this value.
4. Find, in the table, the factor corresponding to the above two intervals, and multiply the maximum velocity by this factor. The result will be the approximate velocity at the time desired.

Current Differences

continued

Location	North Latitude	West Longitude	Slack before Flood	Max Flood	Slack before Ebb	Max Ebb	Speed Ratio Flood	Speed Ratio Ebb	Flood Dir	Ebb Dir	Flood knots	Ebb knots
			h m	h m	h m	h m			Dir	Dir	knots	knots
			ON POLLOCK RIP									
Katama Pt, 0.6 nm NNW of, Katama	41°21.9'	70°30.3'	+0:12	–0:43	+0:20	–0:31	0.3	0.3	325°	180°	0.6	0.5
East Chop–Squash Meadow, betwen	41°27.9'	70°32.2'	+2:07	+0:55	+1:43	+2:04	0.7	1.1	131°	329°	1.4	1.8
East Chop, 1 nm N of	41°29.1'	70°33.5'	+2:40	+1:52	+2:17	+2:11	1.1	1.3	116°	297°	2.2	2.2
West Chop, 0.8 nm N of	41°29.6'	70°35.7'	+2:49	+1:58	+2:20	+2:35	1.6	1.8	096°	282°	3.1	3.0
Hedge Fence–L'Hommedieu Shoal	41°30.3'	70°32.2'	+2:27	+1:38	+2:01	+1:52	1.0	1.3	106°	276°	2.1	2.2
Waquoit Bay entrance	41°32.9'	70°31.8'	+3:21	+2:14	+3:40	+4:01	0.8	0.8	348°	203°	1.5	1.4
L'Hommedieu Shoal, N of W end	41°31.6'	70°34.6'	+2:30	+2:03	+2:12	+2:11	1.2	1.4	080°	268°	2.3	2.3
Nobska Point, 1.8 nm E of	41°31.1'	70°37.1'	+2:13	+1:45	+1:55	+1:49	1.2	1.0	063°	240°	2.3	1.7
VINEYARD SOUND												
West Chop, 0.2 nm W of	41°29.0'	70°36.6'	+1:19	+1:34	+1:50	+1:16	1.3	0.8	059°	241°	2.7	1.4
Nobska Point, 1 nm SE of	41°30.1'	70°38.6'	+2:33	+2:15	+2:25	+2:19	1.3	1.4	071°	259°	2.6	2.4
Norton Point, 0.5 nm N of	41°28.1'	70°39.9'	+1:55	+1:44	+2:01	+1:12	1.7	1.4	050°	240°	3.4	2.4
Tarpaulin Cove, 1.5 nm E of	41°28.3'	70°43.5'	+2:49	+2:07	+2:12	+2:33	1.0	1.4	055°	232°	1.9	2.3
Robinsons Hole, 1.2 nm SE of	41°26.1'	70°46.8'	+2:30	+1:51	+2:11	+2:02	1.0	1.2	060°	240°	1.9	2.1
Gay Head, 3 nm NE of	41°23.1'	70°47.0'	+2:25	+1:50	+1:42	+2:11	0.5	0.8	081°	238°	0.9	1.3
Gay Head, 3 nm N of	41°24.1'	70°51.2'	+2:13	+1:24	+1:55	+1:17	0.6	0.7	074°	255°	1.1	1.2
Gay Head, 1.5 nm NW of	41°21.8'	70°51.8'	+1:30	+0:54	+1:42	+1:16	1.0	1.2	012°	249°	2.0	2.0
VINEYARD SOUND–BUZZARDS BAY												
Woods Hole												
			ON CAPE COD CANAL									
South end Woods Hole	41°30.8'	70°40.2'	+0:29	+1:40	+1:17	+0:08	0.4	0.2	135°	318°	1.5	1.1
0.1 nm SW of Devils Foot I	41°31.2'	70°41.1'	+0:20	+1:41	+0:55	+0:31	0.9	0.8	094°	276°	3.5	3.6
North end Woods Hole	41°31.5'	70°41.6'	–0:29	+1:25	+1:09	–0:04	0.2	0.2	160°	007°	0.8	0.7
Robinsons Hole												
South end Robinsons Hole	41°26.7'	70°48.2'	+1:14	+1:42	+1:20	+1:01	0.2	0.2	162°	339°	0.8	1.0
Middle Robinsons Hole	41°27.0'	70°48.4'	+1:30	+2:00	+1:02	+0:47	0.7	0.6	146°	316°	2.8	2.9
North end Robinsons Hole	41°27.4'	70°48.7'	+1:54	+2:00	+0:52	+1:17	0.2	0.3	161°	338°	1.0	1.2
Quicks Hole												
South end Quicks Hole	41°26.3'	70°50.5'	+2:18	+1:42	+1:17	+0:53	0.5	0.4	140°	300°	1.9	2.0

TABLE FOR FINDING DISTANCE OFF WITH SEXTANT UP TO 6 MILES

Distance in Miles & Cables	HEIGHT OF OBJECT, TOP LINE METERS—LOWER LINE FEET												Distance in Miles & Cables
	12 40	15 50	18 60	21 70	24 80	27 90	30 100	33 110	37 120	40 130	43 140	46 150	
m c	° ′	° ′	° ′	° ′	° ′	° ′	° ′	° ′	° ′	° ′	° ′	° ′	m c
0 1	3 46	4 42	5 38	6 34	7 30	8 25	9 20	10 15	11 10	12 04	12 58	13 52	0 1
0 2	1 53	2 21	2 49	3 18	3 46	4 14	4 42	5 10	5 38	6 06	6 34	7 02	0 2
0 3	1 15	1 34	1 53	2 12	2 31	2 49	3 08	3 27	3 46	4 05	4 23	4 42	0 3
0 4	0 57	1 11	1 25	1 39	1 53	2 07	2 21	2 35	2 49	3 04	3 18	3 32	0 4
0 5	0 45	0 57	1 08	1 19	1 30	1 42	1 53	2 04	2 16	2 27	2 38	2 49	0 5
0 6	0 38	0 47	0 57	1 06	1 15	1 25	1 34	1 44	1 53	2 02	2 12	2 21	0 6
0 7	0 32	0 40	0 48	0 57	1 05	1 13	1 21	1 29	1 37	1 45	1 53	2 01	0 7
0 8	0 28	0 35	0 42	0 49	0 57	1 04	1 11	1 18	1 25	1 32	1 39	1 46	0 8
0 9	0 25	0 31	0 38	0 44	0 50	0 57	1 03	1 09	1 15	1 22	1 28	1 34	0 9
1 0	0 23	0 28	0 34	0 40	0 45	0 51	0 57	1 02	1 08	1 14	1 19	1 25	1 0
1 1	0 21	0 26	0 31	0 36	0 41	0 46	0 51	0 57	1 02	1 07	1 12	1 17	1 1
1 2	0 19	0 24	0 28	0 33	0 38	0 42	0 47	0 52	0 57	1 01	1 06	1 11	1 2
1 3	0 17	0 22	0 26	0 30	0 35	0 39	0 44	0 48	0 52	0 57	1 01	1 05	1 3
1 4	0 16	0 20	0 24	0 28	0 32	0 36	0 40	0 44	0 48	0 53	0 57	1 01	1 4
1 5	0 15	0 19	0 23	0 26	0 30	0 34	0 38	0 41	0 45	0 49	0 53	0 57	1 5
1 6	0 14	0 18	0 21	0 25	0 28	0 32	0 35	0 39	0 42	0 46	0 49	0 53	1 6
1 7	0 13	0 17	0 20	0 23	0 27	0 30	0 33	0 37	0 40	0 43	0 47	0 50	1 7
1 8	0 13	0 16	0 19	0 22	0 25	0 28	0 31	0 35	0 38	0 41	0 44	0 47	1 8
1 9	0 12	0 15	0 18	0 21	0 24	0 27	0 30	0 33	0 36	0 39	0 42	0 45	1 9
2 0	0 11	0 14	0 17	0 20	0 23	0 25	0 28	0 31	0 34	0 37	0 40	0 42	2 0
2 1	0 10	0 14	0 16	0 19	0 22	0 24	0 27	0 30	0 32	0 35	0 38	0 40	2 1
2 2	0 10	0 13	0 15	0 18	0 21	0 23	0 26	0 28	0 31	0 33	0 36	0 39	2 2
2 3	0 10	0 12	0 14	0 17	0 20	0 22	0 25	0 27	0 30	0 32	0 34	0 37	2 3
2 4	0 10	0 12	0 14	0 17	0 19	0 21	0 24	0 26	0 28	0 31	0 33	0 35	2 4
2 5	0 9	0 11	0 13	0 16	0 18	0 20	0 23	0 25	0 27	0 29	0 32	0 34	2 5
2 6	0 9	0 11	0 13	0 15	0 17	0 20	0 22	0 24	0 26	0 28	0 30	0 33	2 6
2 7	0 9	0 10	0 12	0 15	0 17	0 19	0 21	0 23	0 25	0 27	0 29	0 31	2 7
2 8	0 8	0 10	0 12	0 14	0 16	0 18	0 20	0 22	0 24	0 26	0 28	0 30	2 8
2 9	0 8	0 10	0 11	0 14	0 16	0 18	0 20	0 21	0 23	0 25	0 27	0 29	2 9
3 0	0 8	0 9	0 10	0 13	0 15	0 17	0 19	0 21	0 23	0 24	0 26	0 28	3 0
3 2				0 12	0 14	0 16	0 18	0 19	0 21	0 23	0 25	0 27	3 2
3 4				0 12	0 13	0 15	0 17	0 18	0 20	0 22	0 23	0 25	3 4
3 6				0 11	0 13	0 14	0 16	0 17	0 19	0 20	0 22	0 24	3 6
3 8				0 10	0 12	0 13	0 15	0 16	0 18	0 19	0 21	0 22	3 8
4 0				0 10	0 11	0 13	0 14	0 16	0 17	0 18	0 20	0 21	4 0
4 2						0 12	0 14	0 15	0 16	0 17	0 19	0 20	4 2
4 4						0 12	0 13	0 14	0 15	0 17	0 18	0 19	4 4
4 6						0 11	0 13	0 14	0 15	0 16	0 17	0 18	4 6
4 8						0 11	0 12	0 13	0 14	0 15	0 16	0 18	4 8
5 0						0 10	0 11	0 12	0 14	0 15	0 16	0 17	5 0
5 2								0 12	0 13	0 14	0 15	0 16	5 2
5 4								0 12	0 13	0 14	0 15	0 16	5 4
5 6								0 11	0 12	0 13	0 14	0 15	5 6
5 8								0 11	0 12	0 13	0 14	0 15	5 8
6 0								0 10	0 11	0 12	0 13	0 14	6 0

TABLE FOR FINDING DISTANCE OFF WITH SEXTANT UP TO 7 MILES

Distance in Miles & Cables	HEIGHT OF OBJECT, TOP LINE METERS—LOWER LINE FEET												Distance in Miles & Cables
	49 160	52 170	55 180	58 190	61 200	64 210	67 220	70 230	73 240	76 250	79 260	82 270	
m c	° ′	° ′	° ′	° ′	° ′	° ′	° ′	° ′	° ′	° ′	° ′	° ′	m c
0 1	14 45	15 37	16 29	17 21	18 13	19 03	19 54	20 43	21 32	22 21	23 09	23 57	0 1
0 2	7 30	7 58	8 25	8 53	9 20	9 48	10 15	10 43	11 10	11 37	12 04	12 31	0 2
0 3	5 01	5 19	5 38	5 57	6 15	6 34	6 53	7 11	7 30	7 48	8 07	8 25	0 3
0 4	3 46	4 00	4 14	4 28	4 42	4 56	5 10	5 24	5 38	5 52	6 06	6 20	0 4
0 5	3 01	3 12	3 23	3 35	3 46	3 57	4 08	4 20	4 31	4 42	4 53	5 05	0 5
0 6	2 31	2 40	2 49	2 59	3 08	3 18	3 27	3 36	3 46	3 55	4 05	4 14	0 6
0 7	2 09	2 17	2 25	2 33	2 41	2 49	2 58	3 06	3 14	3 22	3 30	3 38	0 7
0 8	1 53	2 00	2 07	2 14	2 21	2 28	2 35	2 42	2 49	2 57	3 04	3 11	0 8
0 9	1 40	1 47	1 53	1 59	2 06	2 12	2 18	2 24	2 31	2 37	2 43	2 49	0 9
1 0	1 30	1 36	1 42	1 47	1 53	1 59	2 04	2 10	2 16	2 21	2 27	2 33	1 0
1 1	1 22	1 27	1 33	1 38	1 43	1 48	1 53	1 58	2 03	2 08	2 14	2 19	1 1
1 2	1 15	1 20	1 25	1 30	1 34	1 39	1 44	1 48	1 53	1 58	2 02	2 07	1 2
1 3	1 10	1 14	1 18	1 23	1 27	1 31	1 36	1 40	1 44	1 49	1 53	1 57	1 3
1 4	1 05	1 09	1 13	1 17	1 21	1 25	1 29	1 33	1 37	1 41	1 45	1 49	1 4
1 5	1 00	1 04	1 8	1 12	1 15	1 19	1 23	1 27	1 30	1 34	1 38	1 42	1 5
1 6	0 57	1 00	1 04	1 07	1 11	1 14	1 18	1 21	1 25	1 28	1 32	1 35	1 6
1 7	0 53	0 57	1 00	1 03	1 07	1 10	1 13	1 16	1 20	1 23	1 26	1 30	1 7
1 8	0 50	0 53	0 57	1 00	1 03	1 06	1 09	1 12	1 15	1 19	1 22	1 25	1 8
1 9	0 48	0 51	0 54	0 57	1 00	1 02	1 05	1 08	1 11	1 14	1 17	1 20	1 9
2 0	0 45	0 48	0 51	0 54	0 57	0 59	1 02	1 05	1 08	1 11	1 14	1 16	2 0
2 1	0 43	0 46	0 48	0 51	0 54	0 57	0 59	1 02	1 05	1 07	1 10	1 13	2 1
2 2	0 41	0 44	0 46	0 49	0 51	0 54	0 57	0 59	1 02	1 04	1 07	1 09	2 2
2 3	0 39	0 42	0 44	0 47	0 49	0 52	0 54	0 57	0 59	1 01	1 04	1 06	2 3
2 4	0 38	0 40	0 42	0 45	0 47	0 49	0 52	0 54	0 57	0 59	1 01	1 04	2 4
2 5	0 36	0 38	0 41	0 43	0 45	0 48	0 50	0 52	0 54	0 57	0 59	1 01	2 5
2 6	0 35	0 37	0 39	0 41	0 44	0 46	0 48	0 50	0 52	0 54	0 57	0 59	2 6
2 7	0 34	0 36	0 38	0 40	0 42	0 44	0 46	0 48	0 50	0 52	0 54	0 57	2 7
2 8	0 32	0 34	0 36	0 38	0 40	0 42	0 44	0 46	0 48	0 50	0 53	0 55	2 8
2 9	0 31	0 33	0 35	0 37	0 39	0 41	0 43	0 45	0 47	0 49	0 51	0 53	2 9
3 0	0 30	0 32	0 34	0 36	0 38	0 40	0 41	0 43	0 45	0 47	0 49	0 51	3 0
3 2	0 28	0 30	0 32	0 34	0 35	0 37	0 39	0 41	0 42	0 44	0 46	0 48	3 2
3 4	0 27	0 28	0 30	0 32	0 33	0 35	0 37	0 38	0 40	0 42	0 43	0 45	3 4
3 6	0 25	0 27	0 28	0 30	0 31	0 33	0 35	0 36	0 38	0 39	0 41	0 42	3 6
3 8	0 24	0 25	0 27	0 28	0 30	0 31	0 33	0 34	0 36	0 37	0 39	0 40	3 8
4 0	0 23	0 24	0 25	0 27	0 28	0 30	0 31	0 33	0 34	0 35	0 37	0 38	4 0
4 2	0 22	0 23	0 24	0 26	0 27	0 28	0 30	0 31	0 32	0 34	0 35	0 36	4 2
4 4	0 21	0 22	0 23	0 24	0 26	0 27	0 28	0 30	0 31	0 32	0 33	0 35	4 4
4 6	0 20	0 21	0 22	0 23	0 25	0 26	0 27	0 28	0 30	0 31	0 32	0 33	4 6
4 8	0 19	0 20	0 21	0 22	0 24	0 25	0 26	0 27	0 28	0 30	0 31	0 32	4 8
5 0	0 18	0 19	0 20	0 21	0 23	0 24	0 25	0 26	0 27	0 28	0 29	0 31	5 0
5 2	0 17	0 18	0 20	0 21	0 22	0 23	0 24	0 25	0 26	0 27	0 28	0 29	5 2
5 4	0 17	0 18	0 19	0 20	0 21	0 22	0 23	0 24	0 25	0 26	0 27	0 28	5 4
5 6	0 16	0 17	0 18	0 19	0 20	0 21	0 22	0 23	0 24	0 25	0 26	0 27	5 6
5 8	0 16	0 17	0 18	0 19	0 19	0 20	0 21	0 22	0 23	0 24	0 25	0 26	5 8
6 0	0 15	0 16	0 17	0 18	0 19	0 20	0 21	0 22	0 23	0 24	0 25	0 25	6 0
6 2					0 18	0 19	0 20	0 21	0 22	0 23	0 24	0 25	6 2
6 4					0 18	0 19	0 20	0 21	0 21	0 22	0 23	0 24	6 4
6 6					0 17	0 18	0 19	0 20	0 21	0 21	0 22	0 23	6 6
6 8					0 17	0 18	0 18	0 19	0 20	0 21	0 22	0 22	6 8
7 0					0 16	0 17	0 18	0 19	0 19	0 20	0 21	0 22	7 0

TABLE FOR FINDING DISTANCE OFF WITH SEXTANT UP TO 7 MILES

Distance in Miles & Cables	HEIGHT OF OBJECT, TOP LINE METERS—LOWER LINE FEET												Distance in Miles & Cables
	85 280	88 290	91 300	94 310	97 320	101 330	104 340	107 350	110 360	113 370	116 380	119 390	
m c	° ′	° ′	° ′	° ′	° ′	° ′	° ′	° ′	° ′	° ′	° ′	° ′	m c
0 1	24 44	25 30	26 16	26 01	27 46	28 29	29 13	29 56	30 38	31 19	32 00	32 41	0 1
0 2	12 58	13 25	13 52	14 08	14 45	15 11	15 37	16 03	16 29	16 55	17 21	17 47	0 2
0 3	8 44	9 02	9 20	9 39	9 57	10 15	10 34	10 52	11 10	11 28	11 46	12 04	0 3
0 4	6 34	6 48	7 02	7 16	7 30	7 44	7 58	8 11	8 25	8 39	8 53	9 07	0 4
0 5	5 16	5 27	5 38	5 49	6 01	6 12	6 23	6 34	6 45	6 56	7 08	7 19	0 5
0 6	4 23	4 33	4 42	4 51	5 01	5 10	5 19	5 29	5 38	5 47	5 47	6 06	0 6
0 7	3 46	3 54	4 02	5 10	4 18	4 26	4 43	4 42	4 50	4 58	5 06	5 14	0 7
0 8	3 18	3 25	3 32	3 39	3 46	3 53	4 00	4 07	4 14	4 21	4 28	4 35	0 8
0 9	2 56	3 02	3 08	3 15	3 21	3 27	3 33	3 40	3 46	3 52	3 58	4 05	0 9
1 0	2 38	2 44	2 49	2 55	3 01	3 06	3 12	3 18	3 23	3 29	3 35	3 40	1 0
1 1	2 24	2 29	2 34	2 39	2 44	2 49	2 55	3 00	3 05	3 10	3 15	3 20	1 1
1 2	2 12	2 17	2 21	2 26	2 31	2 35	2 40	2 45	2 49	2 54	2 59	3 04	1 2
1 3	2 02	2 06	2 10	2 15	2 19	2 23	2 28	2 32	2 36	2 41	2 45	2 49	1 3
1 4	1 53	1 57	2 01	2 05	2 09	2 13	2 17	2 21	2 25	2 29	2 37	2 37	1 4
1 5	1 46	1 49	1 53	1 57	2 01	2 04	2 08	2 12	2 16	2 19	2 23	2 27	1 5
1 6	1 39	1 42	1 46	1 50	1 53	1 57	2 00	2 04	2 07	2 11	2 14	2 18	1 6
1 7	1 33	1 36	1 40	1 43	1 46	1 50	1 53	1 56	2 00	2 03	2 06	2 10	1 7
1 8	1 28	1 31	1 34	1 37	1 40	1 44	1 47	1 50	1 53	1 56	1 59	2 02	1 8
1 9	1 23	1 26	1 29	1 32	1 35	1 38	1 41	1 44	1 47	1 50	1 53	1 56	1 9
2 0	1 19	1 22	1 25	1 28	1 30	1 33	1 36	1 39	1 42	1 45	1 47	1 50	2 0
2 1	1 15	1 18	1 21	1 23	1 26	1 29	1 32	1 34	1 37	1 40	1 42	1 45	2 1
2 2	1 12	1 15	1 17	1 20	1 22	1 25	1 27	1 30	1 33	1 35	1 38	1 40	2 2
2 3	1 09	1 11	1 14	1 16	1 19	1 21	1 24	1 26	1 29	1 31	1 33	1 36	2 3
2 4	1 06	1 08	1 11	1 13	1 15	1 18	1 20	1 22	1 25	1 27	1 30	1 32	2 4
2 5	1 03	1 06	1 08	1 10	1 12	1 15	1 17	1 19	1 21	1 24	1 26	1 28	2 5
2 6	1 01	1 03	1 05	1 07	1 10	1 12	1 14	1 16	1 18	1 20	1 23	1 25	2 6
2 7	0 59	1 01	1 03	1 05	1 07	1 09	1 11	1 13	1 15	1 17	1 20	1 23	2 7
2 8	0 57	0 59	1 01	1 03	1 05	1 07	1 09	1 11	1 13	1 15	1 17	1 19	2 8
2 9	0 55	0 57	0 58	1 00	1 02	1 04	1 06	1 08	1 10	1 12	1 14	1 16	2 9
3 0	0 53	0 55	0 57	0 58	1 00	1 02	1 04	1 06	1 08	1 10	1 12	1 14	3 0
3 2	0 49	0 51	0 53	0 55	0 57	0 58	1 00	1 02	1 04	1 05	1 07	1 09	3 2
3 4	0 47	0 48	0 50	0 52	0 53	0 55	0 57	0 58	1 00	1 02	1 03	1 05	3 4
3 6	0 44	0 46	0 47	0 49	0 50	0 52	0 53	0 55	0 57	0 58	1 00	1 01	3 6
3 8	0 42	0 43	0 45	0 46	0 48	0 49	0 51	0 52	0 54	0 55	0 57	0 58	3 8
4 0	0 40	0 41	0 42	0 44	0 45	0 47	0 48	0 49	0 51	0 52	0 54	0 55	4 0
4 2	0 38	0 39	0 40	0 42	0 43	0 44	0 46	0 47	0 48	0 50	0 51	0 53	4 2
4 4	0 36	0 37	0 39	0 40	0 41	0 42	0 44	0 45	0 46	0 48	0 49	0 50	4 4
4 6	0 34	0 36	0 37	0 38	0 39	0 41	0 42	0 43	0 44	0 45	0 47	0 48	4 6
4 8	0 33	0 34	0 35	0 37	0 38	0 39	0 40	0 41	0 42	0 44	0 45	0 46	4 8
5 0	0 32	0 33	0 34	0 35	0 36	0 37	0 38	0 40	0 41	0 42	0 43	0 44	5 0
5 2	0 30	0 32	0 33	0 34	0 35	0 36	0 37	0 38	0 39	0 40	0 41	0 42	5 2
5 4	0 29	0 30	0 31	0 32	0 34	0 34	0 36	0 37	0 38	0 39	0 40	0 41	5 4
5 6	0 28	0 29	0 30	0 31	0 32	0 33	0 34	0 35	0 36	0 37	0 38	0 39	5 6
5 8	0 27	0 28	0 29	0 30	0 31	0 32	0 33	0 34	0 35	0 36	0 37	0 38	5 8
6 0	0 26	0 27	0 28	0 29	0 30	0 31	0 32	0 33	0 34	0 35	0 36	0 37	6 0
6 2	0 26	0 26	0 27	0 28	0 29	0 30	0 31	0 32	0 33	0 34	0 35	3 06	6 2
6 4	0 25	0 26	0 27	0 27	0 28	0 29	0 30	0 31	0 32	0 33	0 34	0 34	6 4
6 6	0 24	0 25	0 26	0 27	0 27	0 28	0 29	0 30	0 31	0 32	0 33	0 33	6 6
6 8	0 23	0 24	0 25	0 26	0 27	0 27	0 28	0 29	0 30	0 31	0 32	0 32	6 8
7 0	0 23	0 23	0 24	0 25	0 26	0 27	0 37	0 38	0 29	0 30	0 31	0 31	7 0

TABLE FOR FINDING DISTANCE OFF WITH SEXTANT UP TO 7 MILES

Distance in Miles & Cables	HEIGHT OF OBJECT, TOP LINE METERS—LOWER LINE FEET												Distance in Miles & Cables
	122 400	137 450	152 500	168 550	183 600	198 650	213 700	244 800	274 900	305 1000	457 1500	610 2000	
m c	° ′	° ′	° ′	° ′	° ′	° ′	° ′	° ′	° ′	° ′	° ′	° ′	m c
0 1	33 20	36 30	39 26	42 08	44 37								0 1
0 2	18 13	20 18	22 21	24 20	26 16	28 08	29 56	33 20	36 30	39 26			0 2
0 3	12 22	13 52	15 20	16 47	18 13	19 37	21 00	23 41	26 16	28 44			0 3
0 4	9 20	10 29	11 37	12 45	13 52	14 58	16 03	18 13	20 18	22 21			0 4
0 5	7 30	8 25	9 20	10 15	11 10	12 04	12 58	14 45	16 30	18 13	26 15		0 5
0 6	6 15	7 02	7 48	8 34	9 20	10 06	10 52	12 22	13 52	15 20	22 20	28 44	0 6
0 7	5 22	6 02	6 42	7 22	8 01	8 41	9 20	10 39	11 56	13 13	19 25	25 10	0 7
0 8	4 42	5 17	5 52	6 27	7 02	7 37	8 11	9 20	10 29	11 37	17 08	22 21	0 8
0 9	4 11	4 42	5 13	5 44	6 15	6 46	7 17	8 19	9 20	10 21	15 19	20 05	0 9
1 0	3 46	4 14	4 42	5 10	5 38	6 06	6 34	7 30	8 25	9 20	13 51	18 13	1 0
1 1	3 25	3 51	4 17	4 42	5 08	5 33	5 59	6 49	7 40	8 30	12 38	16 39	1 1
1 2	3 08	3 32	3 55	4 19	4 42	5 05	5 29	6 15	7 02	7 48	11 37	15 20	1 2
1 3	2 54	3 16	3 37	3 59	4 20	4 42	5 04	5 47	6 30	7 13	10 45	14 12	1 3
1 4	2 41	3 02	3 22	3 42	4 02	4 22	4 42	5 22	6 02	6 42	10 00	13 13	1 4
1 5	2 31	2 49	3 08	3 27	3 46	4 05	4 23	5 01	5 38	6 15	9 20	12 22	1 5
1 6	2 21	2 39	2 57	3 14	3 32	3 49	4 07	4 42	5 17	5 52	8 46	11 37	1 6
1 7	2 13	2 30	2 46	3 03	3 19	3 36	3 52	4 26	4 59	5 32	8 15	10 57	1 7
1 8	2 06	2 21	2 37	2 53	3 08	3 24	3 40	4 11	4 42	5 13	7 48	10 21	1 8
1 9	1 59	2 14	2 29	2 44	2 58	3 13	3 28	3 58	4 27	4 57	7 25	9 50	1 9
2 0	1 53	2 07	2 21	2 35	2 49	3 04	3 18	3 46	4 14	4 42	7 02	9 20	2 0
2 1	1 48	2 01	2 15	2 28	2 41	2 55	3 08	3 35	4 02	4 29	6 41	8 53	2 1
2 2	1 43	1 56	2 08	2 21	2 34	2 47	3 00	3 25	3 51	4 17	6 23	8 30	2 2
2 3	1 38	1 51	2 03	2 15	2 27	2 40	2 52	3 16	3 41	4 05	6 07	8 09	2 3
2 4	1 34	1 46	1 58	2 10	2 21	2 33	2 45	3 08	3 32	3 55	5 52	7 48	2 4
2 5	1 30	1 42	1 53	2 04	2 16	2 27	2 38	3 01	3 23	3 46	5 38	7 30	2 5
2 6	1 27	1 38	1 49	2 00	2 10	2 21	2 32	2 54	3 16	3 37	5 25	7 13	2 6
2 7	1 24	1 34	1 45	1 55	2 06	2 16	2 27	2 47	3 08	3 29	5 13	6 57	2 7
2 8	1 21	1 31	1 41	1 51	2 01	2 11	2 21	2 41	3 02	3 22	5 02	6 42	2 8
2 9	1 18	1 28	1 37	1 47	1 57	2 07	2 16	2 36	2 55	3 15	4 52	6 28	2 9
3 0	1 15	1 25	1 34	1 44	1 53	2 02	2 12	2 31	2 49	3 08	4 42	6 15	3 0
3 2	1 11	1 20	1 28	1 37	1 46	1 55	2 04	2 21	2 39	2 57	4 24	5 52	3 2
3 4	1 07	1 15	1 23	1 31	1 40	1 48	1 56	2 13	2 30	2 46	4 09	5 32	3 4
3 6	1 03	1 11	1 19	1 26	1 34	1 42	1 50	2 06	2 21	2 37	3 55	5 13	3 6
3 8	1 00	1 07	1 14	1 22	1 29	1 37	1 44	1 59	2 14	2 29	3 43	4 57	3 8
4 0	0 57	1 04	1 11	1 18	1 25	1 32	1 39	1 53	2 07	2 21	3 31	4 42	4 0
4 2	0 54	1 01	1 07	1 14	1 21	1 28	1 34	1 48	2 01	2 15	3 21	4 29	4 2
4 4	0 51	0 58	1 04	1 11	1 17	1 24	1 30	1 43	1 56	2 08	3 12	4 17	4 4
4 6	0 49	0 55	1 01	1 08	1 14	1 20	1 26	1 38	1 51	2 03	3 04	4 05	4 6
4 8	0 47	0 53	0 59	1 05	1 11	1 17	1 22	1 34	1 46	1 58	2 57	3 55	4 8
5 0	0 45	0 51	0 57	1 02	1 08	1 14	1 19	1 30	1 42	1 53	2 50	3 46	5 0
5 2	0 43	0 49	0 54	1 00	1 05	1 11	1 16	1 27	1 38	1 49	2 44	3 38	5 2
5 4	0 42	0 47	0 52	0 58	1 03	1 08	1 13	1 24	1 34	1 45	2 28	3 30	5 4
5 6	0 40	0 45	0 50	0 56	1 01	1 06	1 11	1 21	1 31	1 41	2 32	3 22	5 6
5 8	0 39	0 44	0 49	0 54	0 58	1 03	1 08	1 18	1 28	1 37	2 26	3 15	5 8
6 0	0 38	0 42	0 47	0 52	0 57	1 01	1 06	1 15	1 25	1 34	2 21	3 09	6 0
6 2	0 36	0 41	0 46	1 50	1 55	0 59	1 04	1 13	1 22	1 31	2 16	3 02	6 2
6 4	0 35	0 40	0 44	0 49	0 53	0 57	1 02	1 11	1 20	1 28	2 12	2 57	6 4
6 6	0 34	0 38	0 43	0 47	0 51	0 56	1 00	1 09	1 17	1 26	2 08	2 51	6 6
6 8	0 33	0 37	0 42	0 46	0 50	0 54	0 58	1 07	1 15	1 23	2 04	2 46	6 8
7 0	0 32	0 36	0 40	0 44	0 48	0 53	0 57	1 05	1 13	1 21	2 01	2 42	7 0

TABLE I DISTANCE OF SEA HORIZON IN NAUTICAL MILES

Height in Meters	Height in Feet	Distance in Miles	Height in Meters	Height in Feet	Distance in Miles	Height in Meters	Height in Feet	Distance in Miles	Height in Meters	Height in Feet	Distance in Miles
0.3	1	1.15	4.3	14	4.30	12.2	40	7.27	55	180	15.4
0.6	2	1.62	4.9	16	4.60	12.8	42	7.44	61	200	16.2
0.9	3	1.99	5.5	18	4.87	13.4	44	7.62	73	240	17.8
1.2	4	2.30	6.1	20	5.14	14.0	46	7.79	85	280	19.2
1.5	5	2.57	6.7	22	5.39	14.6	48	7.96	98	320	20.5
1.8	6	2.81	7.3	24	5.62	15.2	50	8.1	110	360	21.8
2.1	7	3.04	7.9	26	5.86	18	60	8.9	122	400	23.0
2.4	8	3.25	8.5	28	6.08	20	70	9.6	137	450	24.3
2.7	9	3.45	9.1	30	6.30	24	80	10.3	152	500	25.7
3.0	10	3.63	9.8	32	6.50	27	90	10.9	183	600	28.1
3.4	11	3.81	10.4	34	6.70	30	100	11.5	213	700	30.4
3.7	12	3.98	11.0	36	6.90	40	130	13.1	244	800	32.5
4.0	13	4.14	11.6	38	7.09	46	150	14.1			

TABLE II DISTANCE OF LIGHTS RISING OR DIPPING

Height of Light		HEIGHT OF EYE												
		Meters												
		1.5	3	4.6	6.1	7.6	9.1	10.7	12.2	13.7	15.2	16.8	18.3	19.8
		Feet												
		5	10	15	20	25	30	35	40	45	50	55	60	65
m	**ft**													
12	**40**	9¾	11	11¾	12½	13	13½	14	14½	15	15½	15¾	16¼	16½
15	**50**	10¾	11¾	12½	13¼	14	14½	15	15½	15¾	16¼	16¾	17	17½
18	**60**	11½	12½	13½	14	14¾	15¼	15¾	16¼	16½	17	17½	17¾	18¼
21	**70**	12¼	13¼	14	14¾	15½	16	16½	17	17¼	17¾	18	18½	19
24	**80**	13	14	14¾	15½	16	16½	17	17½	18	18½	18¾	19¼	19½
27	**90**	13½	14½	15½	16	16¾	17¼	17¾	18¼	18½	19	19½	19¾	20¼
30	**100**	14	15	16	16½	17¼	17¾	18¼	18¾	19¼	19½	20	20½	20¾
34	**110**	14½	15¾	16½	17¼	17¾	18¼	19	19¼	19¾	20¼	20½	21	21¼
37	**120**	15¼	16¼	17	17¾	18¼	19	19½	20	20¼	20¾	21	21½	22
40	**130**	15¾	16¾	17½	18¼	19	19½	20	20½	20¾	21¼	21½	22	22½
43	**140**	16¼	17¼	18	18¾	19½	20	20½	21	21¼	21¾	22	22½	23
46	**150**	16¾	17¾	18½	19¼	19¾	20½	21	21¼	21¾	22¼	22½	23	23¼
49	**160**	17	18¼	19	19¾	20¼	20¾	21½	21¾	22¼	22¾	23	23½	23¾
52	**170**	17½	18½	19½	20	20¾	21¼	21¾	22¼	22¾	23	23½	24	24¼
55	**180**	18	19	20	20½	21¼	21¾	22¼	22¾	23	23½	24	24¼	24¾
58	**190**	18½	19½	20¼	21	21½	22	22¾	23	23½	24	24½	24¾	25
61	**200**	18¾	20	20¾	21½	22	22½	23	23½	24	24½	24¾	25¼	25½
64	**210**	19¼	20¼	21	21¾	22½	23	23½	24	24½	24¾	25¼	25½	26
67	**220**	19½	20¾	21½	22¼	22¾	23¼	24	24¼	24¾	25¼	25½	26	26¼
70	**230**	20	21	22	22½	23¼	23¾	24¼	24¾	25	25½	26	26¼	26¾
73	**240**	20½	21½	22¼	23	23½	24	24½	25	25½	26	26¼	26¾	27
76	**250**	20¾	21¾	22½	23¼	24	24½	25	25½	26	26¼	26¾	27	27½
79	**260**	21	22¼	23	23¾	24¼	24¾	25¼	25¾	26¼	26¾	27	27½	27¾
82	**270**	21½	22½	23¼	24	24½	25¼	25¾	26¼	26½	27	27½	27¾	28¼
85	**280**	21¾	23	23¾	24½	25	25½	26	26½	27	27½	27¾	28	28½
88	**290**	22	23¼	24	24¾	25¼	26	26½	26¾	27¼	27¾	28	28½	28¾
91	**300**	22½	23½	24½	25	25¾	26¼	26¾	27¼	27½	28	28½	28¾	29¼
95	**310**	22¾	24	24¾	25½	26	26½	27	27½	28	28½	28¾	29	29½
98	**320**	23	24¼	25	25¾	26¼	27	27½	27¾	28¼	28¾	29	29½	29¾
100	**330**	23½	24½	25¼	26	26½	27¼	27¾	28	28½	29	29½	29¾	30
104	**340**	23¾	24¾	25¾	26¼	27	27½	28	28½	29	29¼	29¾	30	30½
107	**350**	24	25	26	26¾	27¼	27¾	28¼	28¾	29¼	29½	30	30½	30¾
122	**400**	25½	26½	27½	28	28¾	29¼	29¾	30¼	30¾	31	31½	32	32¼
137	**450**	27	28	28¾	29½	30	30¾	31¼	31¾	32	32½	33	33¼	33¾

(1) No.	(2) Name and location	(3) Position	(4) Characteristic	(5) Height	(6) Range	(7) Structure	(8) Remarks
			MASSACHUSETTS – First District				
	MARTHA'S VINEYARD TO BLOCK ISLAND (Chart 13218)	N/W					
	Quicks Hole						
15910	– *Entrance Lighted Bell Buoy 1*	41 25.8 70 50.4	**Fl G** 4^s		4	Green.	Replaced by can when endangered by ice.
15915	– *Ledge Lighted Buoy 2*		**Fl R** 4^s		3	Red.	Replaced by nun when endangered by ice.
15920	Felix Ledge Buoy 3					Green can.	
	Canapitsit Channel						
15925	– Entrance Bell Buoy CC	41 25.0 70 54.4				Red and white stripes with red spherical topmark.	Removed when endangered by ice.
15930	– Buoy 1					Green can.	
15935	– Buoy 2					Red nun.	
15940	– Buoy 3	41 25.3 70 54.4				Green can.	
15945	– Buoy 5					Green can.	
	Menemsha Creek						
15950	– Entrance Bell Buoy 1	41 21.4 70 46.3				Green.	
15955	– ENTRANCE JETTY LIGHT 3	41 21.3 70 46.1	**Fl G** 4^s	25	9	SG on skeleton tower.	
15960	– Entrance Jetty Daybeacon 2					TR on spindle.	
15965	– Buoy 4					Red nun.	
15970	– Daybeacon 5					SG on pile.	Private aid.
15975	– Daybeacon 6					TR on pile.	Private aid.
	Buzzards Bay Main Channel						
15980 625	*Narragansett-Buzzards Bay Approach Lighted Whistle Buoy A*	41 06.0 71 23.4	**Mo (A) W**		6	Red and white stripes with red spherical topmark.	RACON: N (**—•**).
15985 630	BUZZARDS BAY ENTRANCE LIGHT	41 23.8 71 02.0	**Fl W** 2.5^s	101	7	Tower on red square superstructure, on four black piles, worded BUZZARDS on sides.	Emergency light of reduced intensity when main light is extinguished. HORN: 2 blasts ev 30^s (2^s bl-2^s bl-2^s bl-24^s si). Lighted throughout 24 hours.
15987 632	**Buzzards Bay Temporary Lighted Horn Buoy B (LNB)**	41 23.8 71 05.1	**Fl W** 4^s		9	Red.	Passing light: Mo (A) W HORN: 2 blasts ev 30^s (2^s bl-2^s si-2^s bl-24^s si) RACON: B (**—•••**)
15995	*Buzzards Bay Oceanographic Lighted Buoy*	41 23.9 71 00.7	**Fl Y** 4^s			Yellow.	Private aid.
16000	*Seventeen-foot Ledge Lighted Buoy 1*	41 25.8 71 02.3	**Iso G** 6^s		4	Green.	
16005	*Thirty–foot Spot Lighted Bell Buoy 2*	41 26.1 71 00.7	**Fl R** 2.5^s		4	Red.	
16010	*Hen and Chickens Lighted Gong Buoy 3*	41 27.0 71 01.1	**Fl G** 4^s		4	Green.	
16015	Hen and Chickens Buoy 1 Off south part of ledge.					Green can.	

(1) No.	(2) Name and location	(3) Position	(4) Characteristic	(5) Height	(6) Range	(7) Structure	(8) Remarks
			MASSACHUSETTS – First District				
	MARTHA'S VINEYARD TO BLOCK ISLAND (Chart 13218)	N/W					
	Buzzards Bay Main Channel						
16025	*Coxens Ledge Lighted Bell Buoy 4* West side of shoal.	41 27.0 70 59.3	**Q R**		4	Red.	
16030	*Penikese Lighted Bell Buoy 6*	41 27.9 70 56.7	**Fl R** 4s		4	Red.	
16035	*Mishaum Ledge Lighted Gong Buoy 5* Off southeast side of ledge.	41 29.0 70 57.4	**Fl G** 4s		4	Green.	Removed when endangered by ice.
16040 16731	DUMPLING ROCKS LIGHT 7 On rock.	41 32.3 70 55.3	**Fl G** 6s	52	8	SG on skeleton tower.	
16045	*Wilkes Ledge Lighted Buoy 7*	41 30.5 70 54.6	**Fl G** 2.5s		4	Green.	Replaced by can when endangered by ice.
16050	*Traffic Lighted Gong Buoy 8*	41 29.0 70 53.6	**Fl R** 2.5s		4	Red.	
16055	*Buzzards Bay Midchannel Lighted Bell Buoy BB* East of Wilkes Ledge.	41 30.8 70 50.1	**Mo (A) W**		6	Red and white stripes with red spherical topmark.	Removed when endangered by ice.
16060	*Buzzards Bay Lighted Gong Buoy 10* Marks western edge of shoal.	41 33.1 70 46.7	**Fl R** 4s		4	Red.	Replaced by LIB when endangered by ice.
16070	*Buzzards Bay Lighted Bell Buoy 11*	41 36.5 70 43.6	**Fl G** 4s		4	Green.	Replaced by LIB when endangered by ice.
16075	*Buzzards Bay Obstruction Lighted Buoy 12*	41 37.4 70 42.4	**Q R**		4	Red.	Removed when endangered by ice.
16080	*Buzzards Bay Shoal Lighted Buoy 14*	41 37.6 70 42.0	**Fl R** 2.5s		4	Red.	Removed when endangered by ice.
16081	Bird Island South Shoal Gong Buoy 15	41 39.0 70 42.8				Green.	Replaced by can from Dec. 1 to Apr. 1.
16082	Bird Island Reef Bell Buoy 17					Green.	Maintained from Mar. 15 to Jan. 1.
	Cleveland Ledge Channel						
16085	**Cleveland East Ledge Light**	41 37.9 70 41.7	**Fl W** 10s	74	17	White cylindrical tower and dwelling on red caisson.	HORN: 1 blast ev 15s (2s bl). Lighted throughout 24 hours. Emergency light of reduced intensity when main light is extinguished. RACON: C (–•–•).
16095	– RANGE FRONT LIGHT On southern end of Abiels Ledge.	41 41.6 70 40.5	**Q G**	27		KRW on white tower.	Visible 4° each side of rangeline.
16100	– RANGE REAR LIGHT 3,167 yards, 015° from front light.		**F G**	74		KRW on white skeleton tower.	Visible 1½° each side of rangeline.
16105	– *Lighted Buoy 1*	41 37.8 70 41.9	**Q G**		4	Green.	Replaced by LIB from Jan. 1 to Mar. 15.
16110	– *Lighted Bell Buoy 2* West of shoal.	41 39.9 70 41.8	**Fl R** 4s		4	Red.	Replaced by nun from Jan. 1 to Mar. 15.
16115	*Cleveland Ledge Oceanographic Lighted Buoy*	41 38.6 70 41.0	**Fl Y** 4s			Yellow.	Private aid.

US SAILING Recommended Plotting Symbols

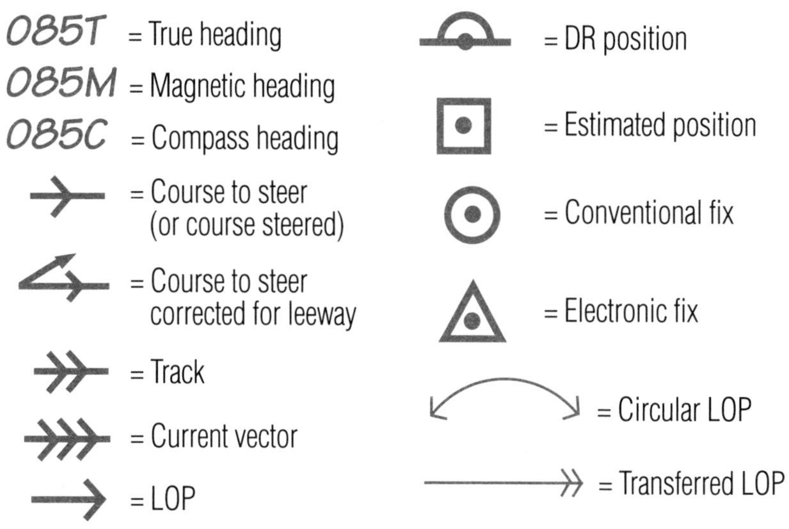

Passage Planning Checklist

- ☐ Review relevant charts and any updates
- ☐ Review relevant updates to Coast Pilot and/or cruising guides
- ☐ Determine distances and estimated times
- ☐ Determine viable alternatives and ports of refuge
- ☐ Identify suitable waypoints and double check their latitude and longitude
- ☐ Check short term weather forecast and developing trends
- ☐ Determine tides and currents and use them to your advantage
- ☐ Identify any tidal height considerations relevant to harbor entrances and exits
- ☐ Identify dangers and obstacles
- ☐ Note approximate basic courses
- ☐ Prepare appropriate foods

Safety Checklist in Fog

- ☐ Double the lookout
- ☐ Hoist the radar reflector
- ☐ Listen hard for sound signals from navigational aids and other vessels
- ☐ Set a radar watch if appropriate
- ☐ Sound your fog signal (— •• under sail, — under power) every two minutes
- ☐ Everyone on board should wear PFDs in case of collision.

Index

Index *continued*

A SPECIAL ACKNOWLEDGMENT TO SAIL AMERICA

Sail America

US SAILING would like to thank SAIL AMERICA for their continuing support of quality sailing instruction and the generous grant they provided to help publish this book.

SAIL AMERICA (originally A.S.A.P.) was formed in 1990 to represent all segments of the sailing industry - from boat builders to sailing schools - with the mission of stimulating public interest in sailing and expanding the sailing market in the U.S.

With this in mind, SAIL AMERICa creates and manages its own boat shows and then reinvests the money earned from the shows to benefit the sport of sailing. SAIL EXPO Atlantic City is now an established annual national show, and in 1995 SAIL EXPO St. Petersburg was successfully launched.

As a result, SAIL AMERICA has been able to reinvest a substantial percentage of show earnings to benefit sailing through several important programs including:

- grants to improve the quality of sailing instruction and training (i.e., funding for this book)
- a grant for the "Class Afloat" program to stimulate awareness of junior and senior high school students in sailing
- grants to provide sailing equipment and promotional materials to associations that benefit physically challenged sailors

SAIL AMERICA's mission is to promote the growth of sailing as a sport, an industry, and a way of life in harmony with the environment. For more information on SAIL AMERICA, please call (401) 841-0900.

NOTES